Polymer Micro- and Nanografting

Polymer Micro- and Nanografting

Celestino Padeste
Paul Scherrer Institut
Department of Synchrotron Radiation and Nanotechnology
Laboratory for Micro- and Nanotechnology
5232 Villigen PSI, Switzerland

Sonja Neuhaus
University of Applied Sciences and Arts Northwestern Switzerland
School of Engineering
Institute of Polymer Nanotechnology
5210 Windisch, Switzerland

AMSTERDAM • BOSTON • HEIDELBERG • LONDON
NEW YORK • OXFORD • PARIS • SAN DIEGO
SAN FRANCISCO • SINGAPORE • SYDNEY • TOKYO
William Andrew is an imprint of Elsevier

William Andrew is an imprint of Elsevier
225 Wyman Street, Waltham, MA 02451, USA
The Boulevard, Langford Lane, Kidlington, Oxford, OX5 1GB, UK

ISBN: 978-0-323-35322-9

Library of Congress Cataloging-in-Publication Data
A catalog record for this book is available from the Library of Congress

British Library Cataloguing-in-Publication Data
A catalogue record for this book is available from the British Library

For Information on all William Andrew publications
visit our website at http://store.elsevier.com/

This book has been manufactured using Print On Demand technology.

Working together
to grow libraries in
developing countries

www.elsevier.com • www.bookaid.org

CONTENTS

Preface: Polymer Structures on Polymer Substrates

Modern synchrotrons are radiation sources built and operated in order to provide high-quality and -intensity photon beams in a very wide range of photon energies. Synchrotron beamlines are usually set up for specific analytical applications ranging from spectroscopic, scattering, and diffraction methods to various types of microscopy and tomography. In addition, a few beamlines have been constructed to perform lithographic structuring, particularly in hard x-ray and more recently in the extreme ultraviolet range. In 2003, we took the unique opportunity to do test exposures on pristine and uncoated fluoropolymer films at the extreme ultraviolet interference (EUV-IL) beamline at the Swiss Light Source (SLS) to create patterns of radicals at their surface. The samples were then treated in a monomer solution at typical conditions to perform graft polymerizations. Analysis of the samples after this reaction indicated that we had a unique method in hand to produce high-resolution patterns of polymer brushes on polymer surfaces. The method appeared attractive because it allowed us to locally endow an inert polymer foil with flexible and functional structures. In the following years, we explored this technology in detail, with emphasis on the influence of reaction parameters on achievable resolution, the formation of polymer brushes and hydrogels, grafting and characterizing polyelectrolyte systems, the production of biofunctional structures, and chemical modification of grafted brush structures in order to introduce additional functionalities.

Furthermore, we widened the scope and potential application fields using different activation methods. Low-pressure and atmospheric-pressure plasma sources as well as vacuum UV lamps were used for large area exposures and, in combination with shadow masks, to produce patterns in the micrometer to millimeter range. Exposures at LIGA beamlines were used to produce bulk structures and electron beam writers to generate arbitrary high-resolution structures.

For this book, we set a broader focus by including work of other groups in related fields. Because there are many reports and reviews available on various aspects of polymeric structures produced by different techniques on a large variety of support materials, we limited the discussion to structures of polymers on polymeric supports produced by top-down structuring combined with grafting technologies. Chapter 1 provides a general introduction on functional polymer structures and polymer brush systems. Structures produced by radiation grafting are summarized in Chapter 2, whereas Chapter 3 is focused on structures produced via immobilization of initiators on polymer surfaces. The range of functionalities achieved in structured polymer-on-polymer systems is described in Chapter 4. Finally, entirely polymeric systems are challenging in terms of the applicability of characterization methods. In Chapter 5, approaches to overcome analytical challenges are presented and illustrated with typical examples.

ACKNOWLEDGEMENTS

Since the start of our work on polymer micro- and nanografting in 2003 numerous students spent their time and effort to explore various aspect of this exciting route to specific functionalization of polymers. Many of their findings are mentioned in this book or reported in the cited publications.

Part of the work included in this book was carried out at different synchrotron beamlines. We would like to acknowledge the possibility to access these great facilities and thank for the support by the teams of XIL and XIL-II, PolLux and TomCat at the Swiss Light Source at PSI as well as LIGA 3 at the Angström Quelle ANKA in Karlsruhe.

Finally we would like to thank our academic partners for interesting and fruitful discussion and Magnus Kristiansen and Matthias Dübner for proofreading the manuscript draft. Thanks go also out to the teams at FHNW and PSI of the joined Institute of Polymer Nanotechnology (INKA), and our partners Roman and Marianne, for their patience and support during the time of manuscript preparation.

CHAPTER *1*

Functional Polymer Structures

Materials with a multitude of functionalities have been introduced in our daily lives and are used without thinking much about their origin and way of production. Examples include magnetic, electrical, optical, and biological functions, and many of them are implemented in polymeric or polymer-based systems. This chapter focuses on the chemical properties and related functionalities that can be introduced to polymer systems using different methods. Main emphasis is placed on polymer brushes, which are extremely versatile and interesting for the functionalization of surfaces and which are accessible on polymers using grafting technology.

1.1 POLYMER SYSTEMS: INERTNESS VERSUS FUNCTIONALITY

Polymers are the most promising materials for current and future applications due to their special properties: They have low densities, exhibit relatively high specific strength and flexibility, and in some cases display remarkable chemical inertness (e.g., fluoropolymers and polyaryletherketones). For many applications, the chemical inertness of polymers provides a substantial advantage. Chemically inert polymers are long-lasting and stable, resistant to weathering, and show very low sorption of water. To benefit from these properties, polymeric coatings are often used as the finish for surfaces of daily life goods, applied in order to achieve (chemical) inertness and stability.

In contrast, other classes of polymers are not inert. They may interact strongly with the environment and adopt special functions. Examples include specific interactions with molecules and ions exploited in separation and purification techniques; electrical and optical properties used in polymer solar cells, organic light emitters, and optical elements; as well as properties relevant for bio-applications, such as anti-biofouling properties and specific binding of proteins.

Polymer Micro- and Nanografting. DOI: http://dx.doi.org/10.1016/B978-0-323-35322-9.00001-2

Adding locally defined functionality to inert surfaces is interesting for many applications. For instance, integration of small sensing elements that could—actively or passively—monitor the freshness of packed goods is of great interest to the packaging industry. With current structuring and patterning technologies, length scales can be addressed which are interesting for studying interactions with cells. Such studies are of importance for the functional design of polymeric implants. Further developments of patterning technology to reach dimensions of the size of single protein molecules are in progress, which will be beneficial for constructing ultrasensitive bioanalytical devices.

In polymer systems, the combination of properties of different components is achieved in a multitude of ways. For example, polymer layers of different functionalities are applied on a polymer substrate by spraying, casting, dip coating, or spreading with a doctor blade. Polymer coatings can be combined with layers of nonpolymeric materials applied from solutions or suspensions of the coating material or precursors thereof, or deposited via vapor phase using physical and chemical vapor deposition.

To achieve structures in deposited layers, screen printing, stamping, and ink-jet printing are very well-established and commercially used technologies. Current and future applications based on such technologies include polymer solar cells [1] and polymer electronics [2]. Polymeric optical waveguide structures on polymer substrates, which require high precision in structure definition, may for instance be produced with micromolding techniques [3]. Taking advantage of the mechanical flexibility of many polymers, developments also aim at elastomeric light-emitting devices and displays [4] or molecularly stretchable electronics [5].

Grafting techniques present an approach on the molecular level for anchoring functionalities on surfaces. The term "grafting" is used in analogy to the biological grafting of a branch from a tree or bush onto the trunk of another closely related species in order to combine the properties of the tree (e.g., the growth of high-quality fruit that in many cases cannot be grown from its own seeds) with the properties of the trunk (e.g., easy to grow and resilient against various insects). In polymer grafting, chemical links are formed between the polymer chains of one material, typically a polymer film, to the chains of another polymer.

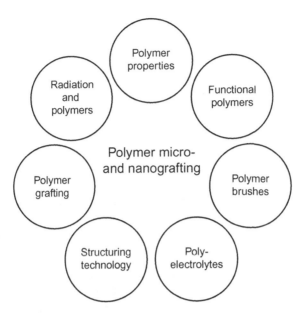

Figure 1.1 Polymer micro- and nanografting in the context of related fields.

In comparison to the methods of layer formation discussed previously, the covalent anchoring of grafted polymer layers intrinsically circumvents adhesion problems that are frequently encountered between different polymers. In addition, grafting techniques offer particular flexibility on the molecular level. A large variety of chemically different monomers and polymers may be grafted, and their strong anchoring allows further functionalization and chemical modification to widen the range of accessible functional properties of the modified polymer surface. Furthermore, polymer chains attached on one end to the surface and grafted in high density usually adopt a so-called polymer brush configuration providing interesting properties as outlined later.

Polymer micro- and nanografting combines grafting techniques with structuring. It is related to a number of fields of science and technologies, as schematically summarized in Figure 1.1. *Polymer materials* provide the substrates for the micro- and nanografting process, which aims at local introduction of specific properties and functionality using *structuring technologies.* Functionality is provided, for instance, by *polymer brushes* or *polyelectrolytes.* Aspects of functional polymer surfaces are summarized in the following section in the context of polymer brushes and are further detailed in Chapter 4. Understanding the impact of *radiation* on polymers in terms of radical formation after ionization and bond-breaking events is fundamental for radiation grafting, which is discussed extensively in Chapter 2. Finally, different aspects of *graft polymerization reactions* are important for successful formation of functional structures starting from either radiation-generated radicals or immobilized initiators, which are discussed in Chapters 2 and 3, respectively.

1.2 POLYMER BRUSHES

Polymer brushes have attracted a tremendous amount of attention in recent years because they enable tailoring of physical, chemical, and biochemical properties of various material surfaces. These brushes are densely packed arrays of polymer chains tethered at one chain end to the surface. At high packing density and in the presence of a good solvent, the chains are forced to elongate perpendicular to the surface. Polymer brushes can be considered as extended interfaces between the polymer surface and the surrounding environment. Due to the flexibility of polymer chains, small molecules are able to penetrate and interact with the chains or with other species embedded within the brush. The term "polymer brush" is not limited to flat surfaces of solid materials; it is also applied to systems in which polymers are densely grafted at the surfaces of nanoparticles, micelles, or even chains of another polymer. The following summary focuses on brushes and brush structures grafted on solid supports.

1.2.1 Formation of Polymer Brushes on Surfaces

Polymer chains may be attached to surfaces by using "grafting-to" or "grafting-from" techniques (Figure 1.2). "Grafting-to" means that preformed polymer chains are bound at one end to the surface or to surface-bound linker molecules via chemical reactions. One intrinsic problem of grafting-to methods is that the bound polymer restricts the diffusion of further chains to the surface, often resulting in low grafting densities. In the "grafting-from" approach, the polymer chains are grown from initiators bound to the surface. Therefore, the term *surface-initiated polymerization* is also used for the same process [6,7]. Typical initiators for free radical polymerizations are azo- and peroxide compounds [8], which are cleaved to reactive radicals under polymerization reaction conditions. In grafting-from processes, only relatively small monomer molecules diffuse to reaction sites at the end

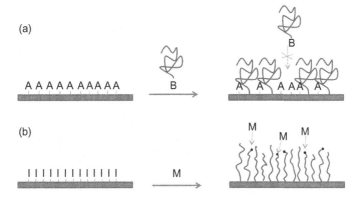

Figure 1.2 **Functional polymer layers on surfaces produced with (a) "grafting-to" and (b) "grafting-from" strategies.** *(a) In* grafting-to *processes, preformed polymer chains containing a linker group (B) are chemically bound to anchor groups (A). (b) In* grafting-from, *polymerization of a monomer (M) is started from initiators (I) bound to the surface.*

of the growing chains; consequently, a higher density of grafted chains can be achieved.

The high polydispersity of grafted chains is one particular drawback of grafting-from techniques based on free radical polymerization to produce polymer brushes. This is often observed and interpreted as a result of poor control over the propagation and termination reactions. In order to circumvent this problem, living radical polymerization schemes such as nitroxide-mediated polymerization (NMP) [9], atom transfer radical polymerization (ATRP) [10,11], reverse ATRP [12,13], or reversible addition–fragmentation chain transfer (RAFT) [12,14,15] have been used, resulting in a significantly lower polydispersity and better control over the chain length.

The previously discussed principles of grafting-to and grafting-from can also be applied for the modification of polymer surfaces with polymer brushes. However, the binding of linkers and polymerization initiators to polymer surfaces is not as straightforward as it is for oxidic inorganic materials. Thus, dedicated pretreatments are usually necessary. These may include rather harsh reaction conditions due to the chemical inertness of many polymers (see Chapter 3). Alternatively, radiation treatment of polymers (to form radicals) followed by exposure to air may be used to form peroxides and hydroperoxides, which can be directly used as initiators for thermally or ultraviolet-induced graft polymerizations [16,17] (see Chapter 2).

1.2.2 Responsive Polymer Brushes

Responsive polymer materials—polymers engineered to change structure and properties in response to environmental influence—are promising materials for biomedical devices, chemical sensors, microfluidic
devices, drug-delivery systems, and self-regulating colloids, as reported
in a series of books and reviews [18,19]. Examples of adaptive properties include surfaces switching between (super)hydrophilic and (super)-
hydrophobic properties upon exposure to different solvents [20], and
photo- or voltage-induced conformational changes of valve-like systems in artificial membranes [21,22]. Polymer brushes are particularly
suitable for the preparation of responsive systems [23,24].

For instance, weak polyelectrolyte brushes (discussed later) are
highly interesting responsive systems because they react with conformational rearrangement to changes in pH and ionic strength [25–27].
As another example, the potential of introducing photoactive moieties
such as azobenzenes or spiropyranes into molecular systems in order
to remotely influence their properties with light has been widely recog-
nized. Under illumination, these molecules undergo (reversible)
changes in configuration, thereby influencing the local chemical environment [28]. Flexible attachment of the photosensitive moieties,
which can be achieved particularly well within polymer brushes, is a
key for their functionality [29]. In mixed polymer brushes or brushes
made up of block copolymers, the responsiveness is often based on the
fact that the characteristics of the layer are dominated by either of the
two components under different conditions [30,31].

1.2.3 Polyelectrolyte Brushes

Polyelectrolytes (PELs) are polymers that carry charges within their
backbone or in side chains. Usually, discrimination is made between
weak and strong PELs. Weak PELs are polymers with weakly acidic or
basic groups, which are protonated or deprotonated depending on the
pH of the surrounding medium, resulting in a pH-dependent charge
density. In contrast, the charge density in strong PELs is not influenced
by the pH.

Polyelectrolyte brushes exhibit interesting characteristics with
respect to both theoretical and practical aspects because their behavior
is fundamentally different from that of uncharged polymer brushes
[32,33]. In the case of strong PEL brushes, in which the charge density

is independent of the pH, the molecular structure and properties are dominated by electrostatic interactions. Mutual repulsion between charged polymer segments strongly influences the physical properties of the grafted layers. In weak PEL brushes—in which the charge density of the chains depends on their protonation level—the chain conformation depends on the pH of the solution. In particular, the swelling of weak PEL brushes in different solvents was extensively studied due to its importance for responsive polymer systems. Swelling depends on the nature of the solvent system, as well as its pH and the concentration and chemical nature of other ions in the solution [34,35]. Furthermore, interactions with selected counterions can be used to tune the wettability of surfaces with anchored PEL brushes [36,37].

1.2.4 Biofunctional Brushes
The discrimination between specific and nonspecific binding on surfaces is a key issue in most biotechnological applications. Polymer brushes have been investigated for binding of proteins via electrostatic interaction, covalent binding, or via metal affinity binding, as well as for the prevention of nonspecific binding, which is often realized with brushes of poly(ethylene glycol) (PEG) derivatives [38].

Due to their extension and flexibility, polymer brushes can accommodate functional proteins at a substantially higher density compared to what is achieved by protein immobilization on flat surfaces. Furthermore, depending on their chemical composition, brushes may provide a mild environment in which the protein is not degraded [39]. In particular, many types of brushes of PELs are interesting candidates for the accommodation of biomolecules because they are swellable in water and provide an ideal chemical environment for maintaining the activity of enzymes and other proteins. The extent of protein binding via electrostatic interaction in PEL brushes is dependent on the sign of charge of the protein and of the PEL, respectively, as well as the ionic strength of the solution [40,41].

1.2.5 Patterned Polymer Brushes
Surfaces that are patterned on the nano- and microscale with responsive polymer brushes are promising for applications in sensing and actuation, as well as for bioanalytical devices [42,43]. Structures of brushes, for instance, are obtained by patterning of preformed brushes using specific etching protocols or by patterning of initiators before the

grafting reaction. Fabrication of patterns by the combination of top-down lithography and surface-initiated polymerizations has been recently reviewed [44]. For instance, photolithography combined with lift-off or etching techniques results in a chemical contrast between exposed and nonexposed areas. This contrast is then transferred into patterns of polymerization initiators [45]. Alternatively, deposited initiator layers can be locally activated or deactivated by lithographic exposures using, for example, a focused electron beam [46–48]. Deposition of polymerization initiators by microcontact printing was also employed to directly deposit patterns of initiators on surfaces [49,50].

To a certain extent, these techniques to obtain structured brushes, which were originally established on solid supports such as silicon, glass, or metal surfaces, can be transferred to polymer surfaces. Such approaches are summarized in Chapter 3. Alternatively, and uniquely accessible on polymer surfaces, locally defined radiation grafting may be applied as discussed in Chapter 2.

REFERENCES

[1] Krebs FC. Fabrication and processing of polymer solar cells: a review of printing and coating techniques. Solar Energy Mater Solar Cells 2009;93(4):394–412.

[2] Barsbay M, Guven O. Grafting in confined spaces: functionalization of nanochannels of track-etched membranes. Radiat Phys Chem 2014;105:26–30.

[3] Ma H, Jen AKY, Dalton LR. Polymer-based optical waveguides: materials, processing, and devices. Adv Mater 2002;14(19):1339–65.

[4] Liang JJ, Li L, Niu XF, Yu ZB, Pei QB. Elastomeric polymer light-emitting devices and displays. Nat Photonics 2013;7(10):817–24.

[5] Savagatrup S, Printz AD, O'Connor TF, Zaretski AV, Lipomi DJ. Molecularly stretchable electronics. Chem Mat 2014;26(10):3028–41.

[6] Edmondson S, Osborne VL, Huck WTS. Polymer brushes via surface-initiated polymerizations. Chem Soc Rev 2004;33(1):14–22.

[7] Barbey R, Lavanant L, Paripovic D, Schuwer N, Sugnaux C, Tugulu S, et al. Polymer brushes via surface-initiated controlled radical polymerization: synthesis, characterization, properties, and applications. Chem Rev 2009;109(11):5437–527.

[8] Roux S, Demoustier-Champagne S. Surface-initiated polymerization from poly(ethylene terephthalate). J Polym Sci Pol Chem 2003;41(9):1347–59.

[9] Li J, Chen XR, Chang YC. Preparation of end-grafted polymer brushes by nitroxide-mediated free radical polymerization of vaporized vinyl monomers. Langmuir 2005;21 (21):9562–7.

[10] Matyjaszewski K, Miller PJ, Shukla N, Immaraporn B, Gelman A, Luokala BB, et al. Polymers at interfaces: using atom transfer radical polymerization in the controlled growth of homopolymers and block copolymers from silicon surfaces in the absence of untethered sacrificial initiator. Macromolecules 1999;32(26):8716–24.

[11] Ramakrishnan A, Dhamodharan R, Ruhe J. Growth of poly(methyl methacrylate) brushes on silicon surfaces by atom transfer radical polymerization. J Polym Sci Pol Chem 2006;44 (5):1758–69.

[12] Yamamoto K, Tanaka H, Sakaguchi M, Shimada S. Well-defined poly(methyl methacrylate) grafted to polyethylene with reverse atom transfer radical polymerization initiated by peroxides. Polymer 2003;44(25):7661–9.

[13] Wang YP, Pei XW, He XY, Lei ZQ. Synthesis and characterization of surface-initiated polymer brush prepared by reverse atom transfer radical polymerization. Eur Polym J 2005;41 (4):737–41.

[14] Yu WH, Kang ET, Neoh KG. Controlled grafting of comb copolymer brushes on poly-(tetrafluoroethylene) films by surface-initiated living radical polymerizations. Langmuir 2005;21(1):450–6.

[15] Mertoglu M, Garnier S, Laschewsky A, Skrabania K, Storsberg J. Stimuli responsive amphiphilic block copolymers for aqueous media synthesised via reversible addition fragmentation chain transfer polymerisation (RAFT). Polymer 2005;46(18):7726–40.

[16] Kang ET, Zhang Y. Surface modification of fluoropolymers via molecular design. Adv Mater 2000;12(20):1481–94.

[17] Chen YJ, Kang ET, Neoh KG, Tan KL. Surface functionalization of poly(tetrafluoroethylene) films via consecutive graft copolymerization with glycidyl methacrylate and aniline. J Phys Chem B 2000;104(39):9171–8.

[18] Minko S. Responsive Polymer Materials: Design and Applications. Ames, IA: Blackwell; 2006.

[19] Chen JK, Chang CJ. Fabrications and applications of stimulus-responsive polymer films and patterns on surfaces: a review. Materials 2014;7(2):805–75.

[20] Xia F, Feng L, Wang ST, Sun TL, Song WL, Jiang WH, et al. Dual-responsive surfaces that switch superhydrophilicity and superhydrophobicity. Adv Mater 2006;18(4):432–6.

[21] Yang B, Yang WT. Novel pore-covering membrane as a full open/close valve. J Membr Sci 2005;258(1–2):133–9.

[22] Tokarev I, Orlov M, Minko S. Responsive polyelectrolyte gel membranes. Adv Mater 2006;18(18):2458–60.

[23] Minko S. Responsive polymer brushes. Polym Rev 2006;46(4):397–420.

[24] Zhou F, Huck WTS. Surface grafted polymer brushes as ideal building blocks for "smart" surfaces. Phys Chem Chem Phys 2006;8(33):3815–23.

[25] Zhou F, Huck WTS. Three-stage switching of surface wetting using phosphate-bearing polymer brushes. Chem Commun 2005;(48):5999–6001.

[26] Zhou F, Shu WM, Welland ME, Huck WTS. Highly reversible and multi-stage cantilever actuation driven by polyelectrolyte brushes. J Am Chem Soc 2006;128(16):5326–7.

[27] Stratakis E, Mateescu A, Barberoglou M, Vamvakaki M, Fotakis C, Anastasiadis SH. From superhydrophobicity and water repellency to superhydrophilicity: smart polymer-functionalized surfaces. Chem Commun 2010;46(23):4136–8.

[28] Piech M, Bell NS. Controlled synthesis of photochromic polymer brushes by atom transfer radical polymerization. Macromolecules 2006;39(3):915–22.

[29] Klajn R, Stoddart JF, Grzybowski BA. Nanoparticles functionalised with reversible molecular and supramolecular switches. Chem Soc Rev 2010;39(6):2203–37.

[30] Lemieux M, Usov D, Minko S, Stamm M, Shulha H, Tsukruk VV. Reorganization of binary polymer brushes: reversible switching of surface microstructures and nanomechanical properties. Macromolecules 2003;36(19):7244–55.

[31] Vyas MK, Schneider K, Nandan B, Stamm M. Switching of friction by binary polymer brushes. Soft Matter 2008;4(5):1024−32.

[32] Ruhe J, Ballauff M, Biesalski M, Dziezok P, Grohn F, Johannsmann D, et al. Polyelectrolyte brushes. Adv Polym Sci 2004;165:79−150.

[33] Ballauff M, Borisov O. Polyelectrolyte brushes. Curr Opin Colloid Interface Sci 2006;11 (6):316−23.

[34] Biesalski M, Ruhe J. Scaling laws for the swelling of neutral and charged polymer brushes in good solvents. Macromolecules 2002;35(2):499−507.

[35] Biesalski M, Johannsmann D, Ruhe J. Synthesis and swelling behavior of a weak polyacid brush. J Chem Phys. 2002;117(10):4988−94.

[36] Moya S, Azzaroni O, Farhan T, Osborne VL, Huck WTS. Locking and unlocking of polyelectrolyte brushes: toward the fabrication of chemically controlled nanoactuators. Angew Chem Int Ed 2005;44(29):4578−81.

[37] Azzaroni O, Brown AA, Huck WTS. Tunable wettability by clicking counterions into polyelectrolyte brushes. Adv Mater 2007;19(1):151−4.

[38] Satomi T, Nagasaki Y, Kobayashi H, Tateishi T, Kataoka K, Otsuka H. Physicochemical characterization of densely packed poly(ethylene glycol) layer for minimizing nonspecific protein adsorption. J Nanosci Nanotechnol 2007;7(7):2394−9.

[39] Hollmann O, Gutberlet T, Czeslik C. Structure and protein binding capacity of a planar PAA brush. Langmuir 2007;23(3):1347−53.

[40] Rosenfeldt S, Wittemann A, Ballauff M, Breininger E, Bolze J, Dingenouts N. Interaction of proteins with spherical polyelectrolyte brushes in solution as studied by small-angle x-ray scattering. Phys Rev E 2004;70(6).

[41] Ladam G, Schaaf P, Cuisinier FJG, Decher G, Voegel JC. Protein adsorption onto auto-assembled polyelectrolyte films. Langmuir 2001;17(3):878−82.

[42] Benetti EM, Zapotoczny S, Vancso J. Tunable thermoresponsive polymeric platforms on gold by "photoiniferter"-based surface grafting. Adv Mater 2007;19(2):268−71.

[43] Konradi R, Ruhe J. Fabrication of chemically microstructured polymer brushes. Langmuir 2006;22(20):8571−5.

[44] Welch ME, Ober CK. Responsive and patterned polymer brushes. J Polym Sci B Polym Phys 2013;51(20):1457−72.

[45] Prucker O, Konradi R, Schimmel M, Habicht J, Rühe J. Photochemical strategies for the preparation and microstructuring of densely grafted polymer brushes. In: Advincula RC, Brittain WJ, Caster KC, Rühe J, editors. Polymer Brushes. Weinheim: Wiley−VCH; 2004. pp. 449−69.

[46] Tsujii Y, Ejaz M, Yamamoto S, Fukuda T, Shigeto K, Mibu K, et al. Fabrication of patterned high-density polymer graft surfaces: II. Amplification of EB-patterned initiator monolayer by surface-initiated atom transfer radical polymerization. Polymer 2002;43(13):3837−41.

[47] Henderson CL, Barstow S, Jeyakumar A, McCoy K, Hess DW, Tolbert LM. Novel approaches to nanopatterning: from surface monolayer initiated polymerization to hybrid organometallic−organic bilayers. Mater Res Soc Symp Proc 2002;705:3−14.

[48] Schmelmer U, Jordan R, Geyer W, Eck W, Golzhauser A, Grunze M, et al. Surface-initiated polymerization on self-assembled monolayers: amplification of patterns on the micrometer and nanometer scale. Angew Chem Int Ed 2003;42(5):559−63.

[49] Shah RR, Merreceyes D, Husemann M, Rees I, Abbott NL, Hawker CJ, et al. Using atom transfer radical polymerization to amplify monolayers of initiators patterned by microcontact printing into polymer brushes for pattern transfer. Macromolecules 2000;33(2):597−605.

[50] Farhan T, Huck WTS. Synthesis of patterned polymer brushes from flexible polymeric films. Eur Polym J 2004;40(8):1599−604.

CHAPTER 2

Polymer-on-Polymer Structures Based on Radiation Grafting

Radiation grafting is a highly versatile method for polymer modification: A huge variety of monomers and monomer mixtures can be grafted from polymeric support materials to endow them with specific properties and functionalities. This chapter summarizes pathways of radiation-induced grafting using photons or electron and ion beams over large energy ranges with a focus on structure formation, either using beam focusing and diffraction or masking techniques. Similarities and differences when using different radiation sources, radiation energies, and types of substrates and monomers to be grafted are discussed as well as achievable structure resolution on surfaces and in the bulk of polymer films.

2.1 INTRODUCTION

Micro- and nanografting is based on one common reaction path summarized in Figure 2.1. In a first step, radiation in the form of photon or particle beams is used to break bonds in a polymer substrate, leading to the formation of radicals. These are stabilized as peroxides or hydroperoxides that are readily formed through reaction with ambient air. The (hydro)peroxides can be split thermally to yield radical initiators for the subsequent graft polymerization. The grafting step is usually carried out in degassed monomer solutions under control of reaction temperature and reaction time. From a chemical standpoint, the reaction is controlled by the kinetics of three basic reactions: initiation, propagation, and termination of chain growth.

Some variations of this scheme are known. For instance, access of air during the exposure process is possible to a certain extent to increase the yield of peroxide species. However, the risk is that the radiation may also induce the formation of ozone, which will then further oxidize the surface rather than leading to peroxide formation. In contrast, contact with air may be avoided completely after exposure, and

Polymer Micro- and Nanografting. DOI: http://dx.doi.org/10.1016/B978-0-323-35322-9.00002-4

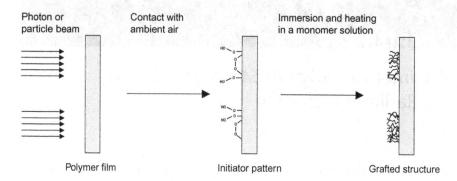

Contact with
ambient air

Immersion and heating
in a monomer solution

Polymer film   Initiator pattern   Grafted structure

*Figure 2.1 **Scheme of the radiation grafting process as described in this chapter.** Polymer films are first exposed with particle or photon beams to break bonds and create radicals at their surface. These react with ambient air to form hydroperoxide or peroxide species. The substrates are immersed into monomer solutions that are then degassed. Upon heating, the peroxides are split to form radicals that initiate the graft polymerization.*

the created radicals may directly be used to initiate polymerization. However, this process needs a more accurate control over the time between exposure and grafting to achieve reproducible results because the radicals formed may be quickly deactivated by recombination or side reactions. Third, the exposure may be carried out with the base polymer immersed in the monomer solution. This simultaneous radiation grafting method has been extensively applied to modify bulk polymers, mainly using γ-radiation. However, it is difficult to apply for structured grafting because shadow masks need to be placed between source and sample. Masks could be placed inside the reaction chamber, requiring a high chemical stability and relatively complicated handling. Alternatively, masks may be placed outside the reactor, which increases the distance between masks and sample, leading to limited resolution.

Based on this very basic process, different approaches are followed to obtain grafted structures on or inside a polymer film. Depending on the characteristics of the radiation—particularly its energy distribution—bonds are cracked preferentially at or near the surface or in the bulk of the polymer. Structure definition in the direction perpendicular to the surface therefore depends on the radiation's penetration depth in the polymer, which is a function of material composition and density, as well as of the energy of the photons or particles. In addition, conditions of the graft reaction can be chosen to some extent to limit the graft reaction mainly to the polymer surfaces, even if the activating beam exposed the whole bulk of the polymer. Lateral structures (parallel to the surface) are achieved by focusing the beams to certain areas

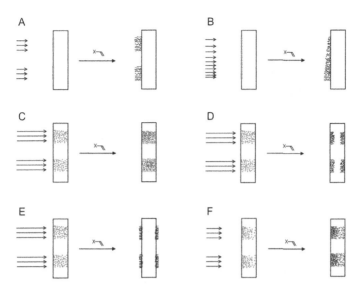

Figure 2.2 **Options for creating structures by radiation grafting exemplified for photon beams of different photon energies.** *(A) Low-energy photons are absorbed and create radicals at or near the surface. Grafting proceeds from the surfaces, leading to brush-like structures if the chain density is high enough. (B) Intensity gradients are transferred into density gradients of the grafted polymer. In the brush regime, the density gradient is reflected as a brush height gradient. (C) High-energy photons penetrate the polymer film, leaving traces of radicals. Graft polymerization from the radicals inside the film leads to bulk structures of graft polymer. (D) As in panel C, but the graft reaction is stopped after a certain reaction time, leaving a partly grafted film with reaction fronts from both sides. (E) As in panel C, but with reaction conditions to favor grafting from the surfaces. (F) Attenuation of medium-energy photon beams leads to a radical density gradient perpendicular to the polymer film, resulting in a gradient in grafting density.*

or by selectively blanking out parts of an unfocused beam using shadow masks. Furthermore, intensity gradients transform directly into gradients in grafting density. Figure 2.2 gives an overview of the main options for structure formation.

2.2 IMPACT OF RADIATION ON POLYMERS

Radiation treatment of polymers is used in many industrial processes and has been investigated in great detail for different types of radiation and polymers [1]. For instance, cross-linking in polymer materials is often achieved by an activation step using ionizing radiation, sometimes followed by thermal treatments to promote the formation of cross-linking bonds. Furthermore, the impact of radiation is of general interest in view of the long-term stability of polymer materials. Much effort has been made to increase stability by improving the resistance to degradation processes, for example, by addition of stabilizers to the polymer material. A review of these aspects would exceed the scope of this book. Therefore, the following discussion focuses on summarizing the most important issues related to radical formation, stabilization, and annihilation, which provide the basis for radiation grafting processes.

2.2.1 Non-ionizing Versus Ionizing Radiation

When considering the impact of radiation on different materials including polymers, it is often useful to discriminate between ionizing and non-ionizing radiation. Non-ionizing radiation mainly excites electrons in chemical bonds. The interaction is therefore strongly dependent on the bond energies. This dependency provides the basis for various analytical methods. For instance, qualitative chemical analyses often probe differences in absorption or emission of specific radiation, and quantitative determination of chemical composition is very often based on quantification of specific absorption or emission intensities. Radical formation upon irradiation with non-ionizing radiation is possible in the case of specific chemical groups that can be activated to form radicals (e.g., benzophenones; see Chapter 3, Section 3.4).

Ionizing radiation provides enough energy to eject electrons from atoms or molecules. The absorption characteristics and primary ionization reactions are therefore mainly dependent on the overall composition and density of the irradiated material. Figure 2.3A schematically depicts the most important events in polymers induced by ionizing radiation. Irradiation may first lead to excitation or ionization within the molecules [2]. In addition to other reaction and relaxation paths, excited molecules may dissociate into two free radicals, whereas the ionized molecules preferentially dissociate into a free radical and a radical ion.

Subsequent reactions are strongly dependent on the chemical nature of the polymer. Recombination of radicals to form a new chemical bond is often observed and is the key process in radiation-induced cross-linking. Examples of polymers in which cross-linking is favored include polyolefins such as polyethylene (PE), natural rubber, or polydimethylsiloxane (PDMS). In other polymers, including most fluorinated polymers, poly(methyl methacrylate) (PMMA), and natural polymers such as DNA and cellulose, chain scission is favored, leading to degradation of the polymer (for a more comprehensive list, see Drobny [2, p. 21]).

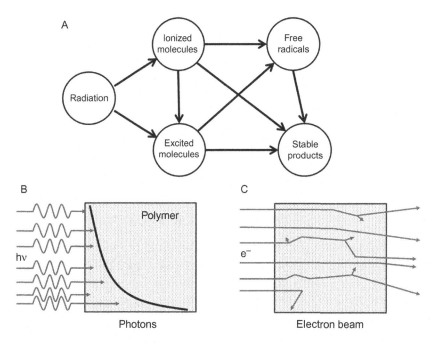

Figure 2.3 **Impact of ionizing radiation on polymers.** *(A) Simplified scheme of radiation-induced events in polymers. (B and C) Comparison of the interaction of UV photons and electrons with polymers. Photon beams are exponentially attenuated along the beam direction, whereas electrons are scattered and slowed down along their path through the polymer film. Source:* Adapted from Drobny [2], with permission from Elsevier Ltd.

For radiation grafting, the stabilization of radicals in the polymer films is very important. Due to the flexibility of polymer chains, rearrangements and chemical interactions are possible over longer distance, in particular above the glass transition temperature. To lower the probability of radical recombination, exposed polymer substrates can be stored at low temperatures to reduce the chain mobility inside the polymer. Temperatures of $-80°C$ are usually sufficient to stabilize the radicals over weeks to months.

2.2.2 Differences and Similarities of Photons and Particle Beams

The most striking difference between the interaction of photons or particles (electrons and ions) and polymers is schematically depicted in Figures 2.3B and 2.3C. Considering a beam where all of the photons have the same energy, absorption of photons in the polymer leads to an attenuation of the beam intensity along the beam direction. The behavior is described by the Lambert−Beer law and is characterized by the attenuation length, which is defined as the depth into the material measured along the direction of the radiation where the intensity of the radiation falls to $1/e$ of its value at the surface. Photons reaching deeper inside the polymer are still of the same photon energy and induce the same type of primary reactions, although at lower density. In contrast, accelerated electrons and ions undergo collisions with atoms along their path in a polymer. They are scattered in different directions, thereby losing energy and slowing down. As a consequence, the impact at some distance below the surface is not the same as that at the surface. In addition, through scattering events, the effective diameter of the particle beam is widened compared to that of the incident beam; this becomes relevant when using shadow masks, for example.

Ionizing particles and photon beams have in common that secondary electrons are created upon their passage through the material. These secondary electrons often still have enough energy to ionize additional atoms. Due to their lower energy compared to that of the primary beam, they reach less far but can travel in all directions. Therefore, highly accelerated ions leave a track in polymer films with orders of magnitude larger diameter than the size of the ions, due to the well-defined reach of secondary electrons (see Section 2.5.3).

For more detailed descriptions of the impact of radiation on polymers, refer to the literature. For example, for photons, see Chapter 4 in Schnabel [3], and for electrons, see Chapter 2 in Drobny [2].

2.2.3 Implications for Shadow Masks
A number of structuring methods discussed here rely on blocking the radiation with a hard mask, which are for instance made of metal. For the absorption of the radiation, basically the same rules apply as those discussed previously for the polymer substrates to be grafted. First, the higher the energy of the radiation, the more mass is needed to block it from the sample. Due to their high densities, high absorption coefficients and good machinability, metals are often the preferred materials. For the production of high-resolution structures with high energy radiation, masks with a high aspect ratio (relation of height to width of the features) are needed (e.g., see Section 2.3.3). However, emission of secondary electrons or induced x-rays from metals exposed with high beam energies can be intense and can lead to a loss in structure definition and resolution.

2.3 RADIATION GRAFTING USING PHOTONS

The following discussion of photon-induced radiation grafting processes is arranged according to the photon energies used. Structure formation using lithographic processes is mostly achieved with radiation in the visible (vis) and ultraviolet (UV) range (down to 193 nm). Photons in this energy range mainly interact with chemical bonds. Materials are often designed and endowed with photosensitive moieties to achieve a specific reaction upon light exposure, such as cross-linking of a polymer. In unmodified polymers, however, the interaction with low-energy radiation is often very weak and radical formation is not a favored process.

In the range above approximately 20–30 eV, the energy of photons is high enough to ionize atoms or molecules in most materials by ejecting electrons from outer shells of atoms or from chemical bonds. At higher energies, electrons from inner shells are also involved. Radicals may then be formed through secondary processes and rearrangements. The absorption of the radiation is strongly material and energy dependent: Figure 2.4 shows the attenuation length of photons in different polymer materials. For low attenuation lengths, the radiation is absorbed close to the surface. The attenuation length generally increases with increasing photon energy, but it is also a function of the elemental composition and the density of the material. Teflon, for instance, has a much shorter attenuation length than polypropylene as a consequence of the fluorine content and the much higher density. Discontinuities in the curves shown in Figure 2.4 correspond to the absorption edges of the corresponding elements (C, N, O, and F).

Because radical formation is related to the absorption of the impinging photons, radiation in the range below approximately 100 eV yields a high density of radicals in surface-near regions of the polymer. As a consequence, surface grafting is achieved when using these radicals as polymerization initiators. When increasing the photon energy—that is, decreasing radiation wavelength—the interaction of the radiation with the polymer is reduced and the transmission through the polymer increases. Therefore, higher energy photons are more suitable to induce bulk grafting.

2.3.1 Visible Light and UV Radiation

As discussed previously, photons in the visible and UV range are not very efficient in breaking sigma bonds and in forming radicals in

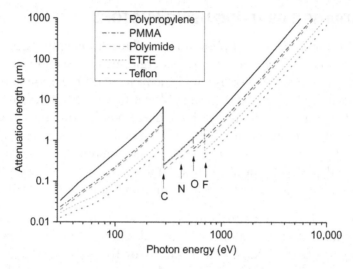

Figure 2.4 Attenuation lengths of x-rays in different polymer materials calculated according to The Center for X-Ray Optics (http:llhenke.lbl.govloptical_constants). The absorption edges of the elements C, N, O, and F are indicated.

polymers. However, photons in this energy range may induce many other chemical reactions, including activation and deactivation of polymerization initiators. Structuring of initiator layers using visible light on various substrates has been thoroughly investigated. These techniques can partly be transferred to initiators bound to polymer surfaces to form polymer-on-polymer structures. Such approaches are reviewed in Chapter 3.

A few cases of direct use of UV light for radiation grafting are reported in the literature. For example, poly(vinylidene fluoride) (PVDF) films were exposed to 297-nm UV irradiation in a nitrogen atmosphere. After exposure to air to form (hydro)peroxides at the surface of the film, PMMA brushes were grafted using free radical polymerization, as confirmed with XPS and IR measurements [4]. An interesting approach to using UV radiation directly for grafting from polymer films was reported using poly(ether ether ketone) (PEEK) as the substrate polymer [5,6]. The diphenylketone moiety in the backbone of PEEK could be activated by UV radiation similar to benzophenone, which is often used as initiator for UV-assisted polymerizations (see Chapter 3, Section 3.4). In a process termed "self-initiated surface grafting," dense and up to 100-nm-thick layers of highly hydrophilic and protein-resistant poly(2-methacryloyl phosphorylcholine) (p-MPC) were grafted from a PEEK surface under UV irradiation using a mercury lamp (see Chapter 4, Section 4.3). The formation of structured surfaces was not demonstrated in any of these reports, but it could be implemented by using shadow masks during the grafting process.

Photons in the vacuum UV region (VUV; typically 100 to 200 nm wavelength) interact with polymers by breaking sigma bonds and forming radicals. VUV radiation can be provided by excimer lamps, which are available at reasonably low cost. Films of PVDF or fluorinated ethylene propylene (FEP) were exposed with VUV radiation of 170 nm wavelength and subsequently grafted with acrylic acid [7]. Grafting was observed not only at the surface but also in the bulk of 50-μm-thick films, which can only be explained by radical formation throughout the film due to surprisingly high transparency of the polymer films in this wavelength range. Bulk-grafted microstructures were obtained by selective exposures through stencil structures or fused silica masks. Our experiments using an Xe_2 excimer lamp to expose 100-μm-thick films of poly(ethylene-*alt*-tetrafluoroethylene) (ETFE) through shadow masks and grafting with glycidyl methacrylate (GMA) confirmed the preferential reaction in the bulk. However, the attenuation of the radiation inside the film resulted in considerably lower grafting levels at the back side of the film compared to the front side.

Lasers as Light Sources
Lasers with very high peak powers and very short photon pulses are available in a wide range of wavelengths—from IR to the UV/vis and VUV range. In the context of polymer processing, they find application particularly in ablation processes in which the high energy density leads to depolymerization or decomposition of the exposed material due to a combination of photochemical and photothermal processes [8]. Laser ablation of polymers is primarily used to obtain three-dimensional structures by using focused beams as writing tools or by exposure through phase masks. Laser ablation was also used to pattern a 100-nm-thick poly(*N*-isopropylacrylamide) (PNIPAAm) layer grafted on polystyrene surfaces with 30- to 50-μm-wide features. Selective ablation of the top 100 nm of the film was used for achieving chemical contrast, which was then used for selective protein adsorption and cell growth studies [9].

When lowering the energy of the laser pulses below the so-called ablation threshold, chemical surface modification starts to dominate over decomposition processes [8]. In the presence of oxygen, the process leads to surface oxidation and is used to locally obtain hydrophilic properties. On such locally oxidized surfaces, initiators might be chemically attached and used for graft polymerizations as described in Chapter 3.

Surface radicals produced in the exposed areas may also serve directly as polymerization initiators. However, their concentration will not be very high because they may be rapidly deactivated at the relatively high temperatures reached locally and because of possible reaction with volatile reaction products. Still, it was reported that poly(hydroxyethyl methacrylate) (PHEMA) could be grafted from PDMS surfaces treated with a CO_2-pulsed laser via the peroxide route [10]. However, the capability of the method to obtain surface structures was not explored in this case.

2.3.2 Structures Via Extreme UV Lithography

Lithographic structuring using extreme ultraviolet (EUV) radiation is currently under intensive development because it is one of the most promising techniques for next-generation lithography for integrated circuit and memory production. Development is currently focused on three main aspects: the development of EUV sources, the development of the optics and masks, and the development of new types of resists. In recent years, dedicated beamlines at synchrotron sources have been built that provide EUV radiation for testing of exposure concepts and resist materials. In addition, a broad range of research using high-resolution nanostructuring is carried out at these beamlines. For the XIL-II beamline at the Swiss Light Source (SLS), a review on these topics was published recently [11]. This beamline is typically operated at 92.5 eV photon energy, which corresponds to 13.5 nm wavelength prioritized for future lithography.

In the context of grafting polymer structures on polymers, the use of EUV radiation is of particular interest because, on the one hand, the photon energy is high enough to efficiently create radicals, and on the other hand, attenuation lengths for 92.5 eV photons are in the range of a few 100 nm for polyolefins and below 100 nm for typical fluoropolymers (Figure 2.4). This means that the photons are absorbed near the surface. Consequently, the radical density will be highest at the surface, and the subsequent graft polymerization will mainly lead to modification of the polymer surface.

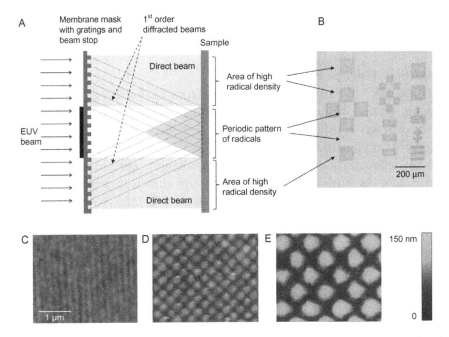

Figure 2.5 **Nano-grafting based on EUV interference lithography.** *(A) Scheme of the EUV-IL exposures performed at the SLS. The incoming beam is diffracted at chromium grating structures on a thin silicon nitride membrane. Interference of first-order diffracted beams yields periodic patterns with twice the frequency of the mask grating. (B) Optical micrograph of a grafted sample containing several test fields. The areas of direct exposure (i.e., high grafting density) are clearly visible. (C–E) AFM images of nanostructures of poly(methacrylic acid) grafted in the interference areas after exposure to two or four interfering EUV beams: (C) line structures with 200-nm period; dot structures with (D) 283-nm and (E) 707-nm period.*

EUV Interference Exposures

Figure 2.5A shows the most often used exposure scheme for lithography testing at SLS. Patterns are achieved using a diffraction mask, produced with e-beam lithography on a thin silicon nitride membrane. The incoming coherent x-ray beam is diffracted by the grating structures. The mask is designed such that first-order diffracted beams overlap in the sample plane, where interference leads to a periodic intensity distribution with half of the period of the grating structures of the mask. Interference of two beams leads to the formation of line structures, and with four beams, dot or hole structures are created. More complicated schemes using three or six beams for structures with hexagonal symmetry and even five or eight beams leading to quasi-crystalline arrangements have also been explored [12].

For studies on graft processes, the undiffracted beam impinging directly on the surfaces is also of high relevance. The resulting homogeneously grafted areas are used to study the dose-height dependence of grafted structures using atomic force microscopy (AFM) or profilometry.

Growth of Polymer Brushes

For a variety of monomers, the grafting from EUV-exposed polymers—mainly from ETFE—has been studied in detail. A number of fields were exposed with increasing dose on a single sample, and the height of the grafted structures was determined in directly exposed fields. Some characteristic examples of dose-height curves, additionally dependent on further graft parameters, are shown in Figure 2.6. The structure height measured in the dry state increases with increasing EUV dose. Most often, the increase in height roughly follows a square root dependence of the applied dose, which is an indication of the formation of brush-like polymer layers [13]. At higher doses, the structure

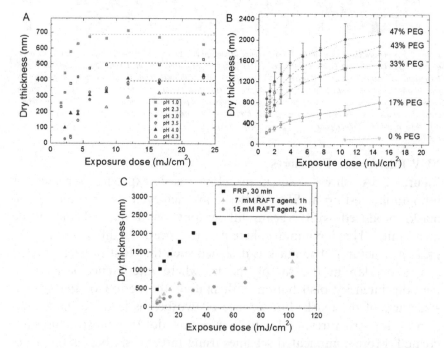

*Figure 2.6 **Control of the graft polymerization as determined with step-height measurements of the brush dry thickness.** ETFE substrates were activated with EUV interference lithography. In most cases, the data points roughly follow a square root dependence on the exposure dose. The brush thickness was also influenced by (A) the pH of the graft solution (aqueous solution of methacrylic acid), (B) the viscosity of the solution (GMA in dioxane), controlled by addition of PEG, and (C) the addition of a RAFT agent to the monomer solution (GMA in methylethyl ketone). For details, see the text. Source: Figure compiled from Neuhaus et al. [14] and Farquet et al. [15,16] with permission from Elsevier Ltd. and ACS.*

height often levels off or even decreases due to the higher probability of radical recombination.

One general finding is that the grafting process is very sensitive: Only very low doses had to be applied compared to resist-based EUV lithography to achieve well-defined structures. This can be explained by two facts. First, the creation of radicals in fluoropolymers and the initiation of the polymerization appear to be very efficient. Second, and probably more important, the grafting reaction provides an intrinsic chemical amplification process; that is, from every radical created, a long polymer chain consisting of hundreds of monomer units can be grown.

Different parameters may be used to control the graft reaction, depending on the type of monomer used. Generally, structure heights increase with higher monomer concentrations (usually between 10 and 100%) and grafting times (typically 20 min to 2 hr). Temperature may also be used as a control, but a temperature of at least 50°C is needed to start the polymerization, whereas above 80−90°C peroxide decomposition and radical recombination are presumably too fast to achieve reproducible results.

Three different means to impose additional control over the graft reaction are exemplified in the graphs of Figure 2.6. First, if the monomer is a weak acid, the pH of the grafting solution has a dramatic impact (Figure 2.6A). At high pH, the growing chains as well as the monomer molecules are deprotonated and carry negative charges, which lowers access of the monomer to the reaction site due to electrostatic repulsion and due to the presence of a hydration sphere around the monomer [14]. Second, increasing the viscosity of the monomer solution—for example, by adding poly(ethylene glycol) (PEG; molecular weight, $\sim$400 g/mol)—was found to strongly increase the grafting yield, most likely due to a decrease in termination rates (Figure 2.6B). Beyond a certain viscosity limit (for data, see Farquet et al. [15]), a decrease in grafting density was observed, arguably because the initiation rate was also lowered under these conditions. Third, starting from peroxide initiators, reversible addition fragmentation chain transfer (RAFT) reaction schemes could be applied (Figure 2.6C). RAFT is an example for controlled living radical polymerizations (see Chapter 3, Section 3.3). RAFT polymerizations are based on the equilibrium between dormant ("capped") and active ("living") radicals. To establish a stable equilibrium and enable sufficient polymerization rates in the case of surface grafting, growing chains are also required in the solution. These were induced and controlled by adding a soluble initiator at defined concentration. It was found that initiator addition boosts the grafting height drastically, probably because the polymerization in the solution increases the viscosity, which in turn increases the polymerization rate at the surface, as discussed previously. Furthermore, growing chains in the solution may bind to growing chains at the surface, thus leading to a combined grafting-to/grafting-from mechanism. Adding the RAFT agent lowers the polymerization rate to levels below the rate observed in free radical polymerization. Because termination is greatly suppressed compared to free radical polymerization, the reaction does not stop as long as monomer remains in the solution. Therefore, the reaction time becomes a more interesting parameter for controlling the brush height. In addition, it was observed that the range of steady increase of structure height with dose was significantly extended toward higher dose because the added RAFT agent prevents the recombination of radicals and of growing chains at the surface.

Structure Formation

The exposures performed in our studies were usually done with masks that provide a number of fields with line and dot patterns of different periodicities. This allows the determination of resolution limitations simultaneously for specific grafting conditions and different exposure doses on one sample. Figures 2.5C–E show typical examples of fields grafted after exposure with two- and four-beam interference. Homogeneous periodic structures are achieved on fields with a size of up to several 100 μm, depending on the mask design. Because the local EUV dose level in the interference region is much lower than that in the directly exposed areas, the height of resulting structures is also usually lower, from 10–100 nm for high-resolution structures to 100–200 nm for lower resolution structures.

At low-exposure doses, grafting results in very shallow profiles, whereas at too high doses the structures combine into a continuous film. In the intermediate range, structure resolution of 200-nm period is routinely obtained for most of the monomers tested. Higher resolution is difficult to achieve, even though the EUV-IL technique allows approximately 10-fold better resolution. The main reason is the growth of relatively long polymer chains with a large polydispersity that start overlapping. In addition, the intensity profile in the interference region is sinusoidal, which translates into relatively shallow profiles with limited structure definition due to the weak dose-height dependence. Improvement of the structure quality was achieved by annealing steps in a good solvent or by vacuum treatment at elevated temperatures, which allowed the grafted polymer chains to rearrange. Furthermore, good resolution of 100-nm period line structures was demonstrated using the previously mentioned RAFT polymerization scheme, which significantly lowers the polydispersity and reduces the polymerization rate in general [15].

In addition to the formation of high-resolution structures, control of the exposure dose allows formation of surface relief structures [17]. A gradient in exposure dose leads to a gradient in grafting density in the subsequent grafting process and, consequently, a gradient in structure height. Furthermore, exposing different fields with different doses allows the production of multilevel structures in a single grafting step. Using a conventional resist-based lithographic approach, such more complex surface structures are difficult to produce and usually involve several sequential structuring steps.

In summary, EUV activation—although currently strongly dependent on access to a dedicated synchrotron beamline—allows well-controlled grafting of polymer brushes from polymer surfaces with high spatial resolution capabilities. The availability of a large variety of monomers suitable for this grafting process, combined with post-polymerization modification processes, opens a wide field to introduce surface functionalities. Select examples are discussed in Chapter 4.

2.3.3 Photons in the keV Range

Photons in this range are mainly accessible from x-ray tubes and synchrotrons. Whereas x-ray tubes deliver characteristic radiation, depending on the cathode material and the applied voltage, the wavelength in synchrotron beamlines can be selected in a wide range. A further advantage of synchrotrons is the much higher flux, leading to much shorter exposures of the films to be grafted. Of particular interest for structured grafting are LIGA beamlines at synchrotrons, which are designed for large-area exposures to create high aspect ratio microstructures (for a review of LIGA, see Malek and Saile [18]).

In 1976, Chapiro *et al.* [19] reported on the grafting of so-called mosaic membranes. In two successive grafting steps, they produced well-localized domains of acrylic acid (AA) and 4-vinylpyridine (4VP) grafted in the bulk of 20- or 50-μm-thick PTFE films. The films were first exposed using an x-ray tube operated at 45 kV. To selectively block the radiation, a 5-mm-thick metal grid mask with 0.5-mm-wide bars and 0.5-mm spacings was placed on the films during the exposure. After grafting with AA, a second x-ray exposure through a mask protecting the grafted regions was performed, followed by a second graft reaction using 4VP. Domains of 0.5 mm in width were clearly visible when selectively stained. The ion pair diffusion of the membranes was 100 to 1000 times higher compared to that of unstructured AA-grafted films, indicating preferential diffusion of cations and anions, respectively, through the chemically different domains in the film.

Our work on bulk structuring was motivated by possible application of micrografted films as fuel cell membranes [20,21]. As detailed in the section on electron-based processes, radiation grafting using monomers such as styrene followed by sulfonation is frequently used to endow high-performance polymers such as ETFE or FEP with ion-conducting properties. However, in the grafting and the chemical modification processes, the substrate materials partially lose their advantageous properties, particularly in terms of mechanical stability. Structured grafting in this case aims at sustaining a stabilizing network of unmodified base polymer in the modified membrane. An overview of typical structures obtained using different photon sources and masks is given in Figure 2.7.

When using relatively soft x-rays from a chromium x-ray tube, considerable attenuation of the radiation in the film needs to be taken into account. The sample shown in Figure 2.7A was only partially grafted; that is, the graft reaction was stopped before the reaction fronts from the front and the backside could join inside the membrane. The graft level at the backside was found to be considerably lower, which can be explained by the attenuation of the x-ray beam to approximately 30% of its initial intensity at the backside of the film. Using higher energy photons of a tungsten x-ray tube operated at 40 kV, well-defined structures throughout the entire film thickness were produced. However, the metal mesh used as the mask led to a high fraction of unmodified polymer, which would lead to a greatly reduced active area when used in a fuel cell. To optimize the ratio between grafted and ungrafted areas, high aspect ratio nickel grid structures were produced by electroplating in a structured polymer film and used as the mask during

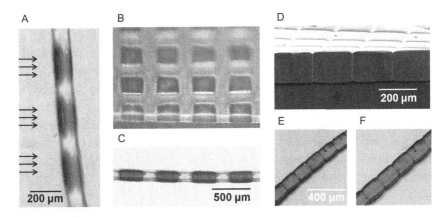

Figure 2.7 **Examples of polystyrene structures grafted in 100-μm-thick ETFE foils after exposure to photons in the keV range.** *All samples were sulfonated and selectively stained to enhance the microscopy contrast. (A) Cross section of a film exposed through wire grid using a Cr x-ray tube. Arrows indicate the direction of the incident photon irradiation. The graft reaction was stopped before the reaction fronts from both sides joined inside the film. (B and C) Top view and cross section of a film exposed with a tungsten x-ray tube. Homogeneous grafting throughout the film was achieved. (D) SEM image of a cut film exposed at a LIGA beamline using a high aspect ratio nickel grid as the mask. The structure is well-defined throughout the film. (E and F) Optical images of the same film in the dry and in the swollen (wet) state.*

exposure at a LIGA beamline [21]. The masking effect of the 15-μm-wide and 100-μm-high nickel bars is evident in the SEM image (Figure 2.7D). To establish the ionic conductivity needed in the fuel cell application, membranes must be soaked with water, leading to swelling of the membrane (Figures 2.7E and 2.7F). Water uptake and membrane swelling is strongly related to the degree of grafting, which can be adjusted by selection of the exposure dose and grafting time. The structured membranes were found to have similar degrees of grafting as the nonstructured ones and to provide slightly lower ionic conductivity due to the lower active area, but they had a significantly longer lifetime when operated in a fuel cell [21].

2.3.4 Gamma Radiation

Gamma radiation is usually defined as radiation with photon energies above 200 keV. Radiation grafting using γ-radiation—for example, from ^{60}Co sources—is a relatively old technique that was investigated, for instance, for the production of graft membranes for fuel cell applications using the simultaneous grafting approach [22]. The radioactive sources are advantageous in that they are independent of electrical equipment; in turn, due to the high photon energy and the concomitant risks, an extremely high level of protective measures is required for safety reasons. In addition, radiation of such high energy shows very high penetration through any material and particularly through polymers. Most of the radiation will therefore not interact with the sample. As a result, long exposure times are required to induce the desired effect, and processes are generally inefficient. Furthermore, the absorption of radiation by materials used to produce masks is rather low at such high photon energies. Therefore, the thickness of absorption masks for structured grafting using γ-radiation would be unreasonably high.

2.4 RADIATION GRAFTING USING ELECTRONS

As in the case of photons, electrons in a wide range of energies are available from different sources. Again, with increasing energy, their impact on polymers shifts from the surface to the bulk. Consequently, depending on the energy of the electrons used for activation, grafting reactions can be confined to the surface or extend into the bulk of the material. An intrinsic difference from photons is the charge of electrons, which may lead to charge accumulation in the material to be modified. This effect can, for instance, lower the achievable resolution of structuring methods. In the following discussion, examples for structures produced in the lower range of electron energies using e-beam lithographic equipment and large-area exposure tools are given, followed by a summary of applications of high-energy electrons. Note that all the electron beams described in the following provide enough energy per electron to ionize atoms in the polymer.

2.4.1 Structures Via Electron Beam Lithography

Among the techniques for producing arbitrary structures at high resolution, sequential writing using focused electron beams is by far the most advanced and best established method. Electrons can well be accelerated and focused using electrostatic and magnetic lenses. The disadvantage of long process times due to serial writing in e-beam lithography is partly overcome by steadily increasing writing speeds and by parallelization of multiple e-beams writing simultaneously on the same wafer.

The main reason why structured polymer-on-polymer grafting based on e-beam lithography has not been studied in detail is the low electrical conductivity of the base polymers of interest. To reach the high performance and resolution of e-beam lithography, at least some conductivity of the substrates is needed to equalize the charge produced during the exposure process. Nonconductive polymers such as PMMA are commonly used as resist materials for e-beam lithography, but in this case the underlying wafer substrates, often covered with a metal layer, provide the required conductivity. As a further limitation, high-resolution e-beam lithography is optimized for perfectly polished wafer surfaces, which provide a flatness that is difficult to achieve using polymer foils.

Despite the intrinsic problems faced when exposing highly insulating polymer films, we recently started experiments with 100-μm-thick ETFE foils in order to explore the possibilities and limits of the technology [23]. Line structures of poly(dimethylaminoethyl methacrylate) (PDMAEMA) were grafted after exposure with two different e-beam lithographic tools—one operated with an acceleration voltage of 2.5 kV and the other operated with an acceleration voltage of 100 kV. AFM was used to investigate the influence of the different exposure parameters on the structures. Well-defined line structures of up to several hundred nanometers in height were grafted from surfaces exposed at low and at high acceleration voltages (Figure 2.8). The process appeared to be very sensitive, similar to the situation described for EUV exposures (see Section 2.3.2). Compared to commonly used resists in e-beam lithography, very low exposure doses were needed to obtain the best structure definition and maximum height of the grafted

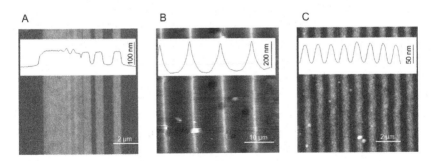

Figure 2.8 AFM images and corresponding line profiles of PDMAEMA brush structures grafted from ETFE supports after defining radical patterns with an e-beam writer. (A) Lines grafted after writing with 2.5 keV electrons show almost rectangular line profiles. The whole structure consists of 10 written lines that were partly overlapping. (B) Lines written with 100 keV are well resolved but show pronounced tailing on both sides. (C) At narrow distances, the tails overlap to a continuous background on which the structures are still well resolved.

structures. The resolution limit with both exposure tools appeared to be in the range of 300 nm—that is, far above the nominal focus achievable by the e-beam systems. This may be due to widening of the beam focus by the accumulated charge at the surface.

Distinct differences were found in the line profiles achieved with the two exposure systems. Using the 2.5-keV beam, grafted lines showed profiles of almost rectangular shape (Figure 2.8A). Using the 100-keV beam resulted in better defined lines with a profile with a clear maximum but with a very broad tailing, which is assigned to a broader impact range of electrons resulting from scattering cascades and emission of secondary electrons. In addition, the higher acceleration voltage leads to a penetration depth of the electrons in the micrometer range. Therefore, considerable grafting in the subsurface region of the film has to be assumed, as also indicated by substantial bulging of the substrates observed after the grafting process.

These test experiments demonstrated the resolution capabilities of e-beam lithographic exposures of highly insulating substrates, but they also revealed some difficulties. Deviations in the micrometer range between the programmed and actual writing position were evident in many instances. Furthermore, at high exposure doses, the written structures appeared distorted, probably resulting from increased accumulation of charge at the surface.

2.4.2 Absorption Mask Techniques Using Low-Energy Electron Beams

The recent development of e-beam sources operated in the range of approximately 80–200 kV opens new fields for polymer (surface) modification. Electrons are generated inside a sealed and evacuated emitter tube and accelerated toward a thin foil acting as a separator to the sample chamber that is maintained at atmospheric pressure under inert gas. The accelerated electrons penetrate the metal foil and can act on samples passing underneath the beam shower in the chamber [2]. The penetration depth of the electrons in this energy range is inversely dependent on the density of the material and reaches 10–20 cm in gases and tens to hundreds of micrometers in polymers [24].

In test experiments, 100-μm-thick ETFE foils were exposed and then grafted for 1 hr with GMA. A 200-μm-thick silicon wafer with 100-μm-sized squared openings was used as the mask. Due to the fabrication process using anisotropic etching, the angle of the opening walls was 55°; that is, the thickness of the mask gradually increased at the edges of the openings. The acceleration voltage was set to different values from 80 to 150 kV, whereas the nominal dose was 25 kGy in all cases. Note that for practical reasons, a 20-μm-thick PET foil had to be placed over the sample and mask. This results in attenuation of the beam, which is more pronounced at lower electron energy; that is, the accumulated dose in the sample was lower at lower acceleration voltage.

From the images of prepared samples presented in Figure 2.9, a strong dependence of the grafting reaction on the electron energy is evident. At 80 kV, the grafted structures are well-defined but very shallow. With increasing electron energy, the penetration depth and the dose accumulated in the sample increase, and the grafting becomes more pronounced. Due to the electron density gradient arising from the slope in mask thickness, the grafted square structures also have

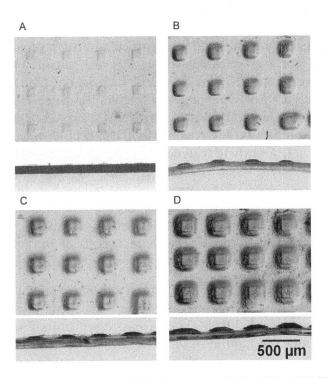

Figure 2.9 **Top view and cross sections of poly-GMA microstructures grafted from 100-μm-thick ETFE foils after e-beam exposure through a microstructured silicon wafer.** *The acceleration voltage was increased in the following sequence: A = 80 kV, B = 100 kV, C = 120 kV, and D = 150 kV. The dose was set to 25 kGy. The grafted regions were selectively stained with a dye reacting with the epoxide of the poly-GMA.*

increasingly slanted side walls with increasing electron energy. The cross sections of the samples show that grafting was not limited to the surface but also reached the bulk of the film. For the sample exposed at 150 kV, even grafting from the backside of the film is evident, which indicates that electrons were partly penetrating the 100-μm fluoropolymer film.

These preliminary results demonstrate the great potential of combining large-area e-beam exposures with shadow masks to create micrografted polymer films. Structure definition in the micrometer range appears feasible because there is room for optimization of materials and geometries of the masks used.

2.4.3 High-Energy Electrons

Electron beam-induced grafting is probably technologically the most important process to modify the chemical composition and, consequently, the chemical and mechanical bulk properties of preformed polymer films. This is of interest for producing membranes such as ion-exchange membranes [22] and particularly for proton-conductive membranes for fuel cell applications [25]. To obtain homogeneous grafting throughout the film thickness, high-energy radiation is preferred, which is hardly attenuated by the film. In the case of electrons, acceleration voltages in the range above 1 MV are typically used. This has a major impact on the possibilities for structure formation. Thick masks of highly absorbing materials are needed to efficiently block the incoming radiation. When aiming at grafted structures in the micrometer range, high aspect ratio structures are required in the mask. In addition, the formation of secondary electrons or photons from the irradiated masks cannot be neglected when using such high-energy electrons. These may create radicals in regions of the polymer film blanked off by the mask, leading to poor structure definition.

The possibility to produce structures was demonstrated by placing a wire grid consisting of 300-μm-thick wires on a 25-μm-thick FEP film during exposure with 1.05 MeV electrons. After grafting with styrene and sulfonation, an imprint of the structure was evident. However, no sharp borders between grafted and nongrafted regions were found. Furthermore, the retention of the flexibility of the original FEP film, which was expected from the structuring process, could not be confirmed experimentally [26].

2.5 RADIATION GRAFTING USING PARTICLE BEAMS

This section discusses the application of ionized and accelerated ions or molecules to generate initiators in polymers for graft polymerizations. In contrast to electrons and photons, the chemical properties of the impinging ions may also play a role with respect to the interaction with the substrate material. For instance, dopant atoms such as phosphorus are implanted in silicon materials using ion accelerators to tune the semiconducting properties of the substrates. For polymer grafting, implanted ions might lead to unwanted side effects. Therefore, noble gas ions are often preferred for radical generation. However, at very high energies, the accelerated ions are not implanted but, rather, penetrate the whole bulk of the polymer substrate. In this case, the chemical nature of the penetrating ion is irrelevant for the properties of the modified material, and only the transfer of energy to break bonds, and to create radicals and cascades of secondary electrons needs to be considered.

As a special source of particles, low-temperature plasmas are included in this section, which strictly speaking are sources of ionized atoms or molecules as well as photons and electrons that all may interact with the polymer. In addition to well-established low-pressure plasma sources, corona discharge or recently developed atmospheric pressure plasma sources may be used for surface modification.

2.5.1 Plasma Activation

Nonthermal plasmas are frequently used to treat polymer surfaces in order to tailor their surface properties, for instance, for bio-applications [27]. In low-pressure plasma chambers, a gas is ionized in an electric field and the formed species—that is, ions, electrons, and photons—will interact with the surface of a polymer substrate placed in the chamber. If an inert gas such as argon or helium is used to generate the plasma, their ions will not form any stable species at the surface. Bond breaking at the surface of the exposed polymer is favored, leading to polymer fragmentation and radical formation. The radicals may recombine or undergo further reactions inside the polymer, but often they are stable enough to be used as graft polymerization initiators. At high power, substantial sputtering of the surface may occur. In the presence of a reactive gas, additional reactions may lead to other surface species. For instance, surface oxidation occurs in an oxygen plasma while amines may be formed at the polymer surface in an ammonia plasma. On such oxidized or amine-terminated surfaces, initiators for graft polymerizations may be bound chemically and used for graft polymerizations (see Chapter 3). Furthermore, monomers can be introduced into the plasma chamber and directly polymerized at the surface.

The effect of plasma treatment on polymer surfaces is often non-permanent because the surfaces can undergo hydrophobic recovery. This means that the activated surface has the tendency to rearrange due to sufficient mobility of the polymer chains, thereby burying the more polar functionalities in the bulk of the polymer. When measuring water contact angles on hydrophobic polymers hydrophilized by plasma, an increase in contact angle can be observed as a function of time. This hydrophobic recovery is strongly dependent on the polymer and on the temperature because it involves temperature-dependent processes such as molecular diffusion and reorganization [28]. Often, the hydrophobic properties are recovered within hours to days. Grafting from plasma-activated surfaces provides longer term stability of surface properties because of the much larger size of grafted chains compared to mere surface functional groups, which limits surface rearrangements.

Grafting of AA on polymer films after activation with low-pressure argon plasma was compared to grafting after γ-irradiation. Despite the lower overall graft level, the grafting density at the surface was higher after plasma activation, due to the preferential radical formation at the polymer surface [29]. A similar reaction has been demonstrated on a poly(ethylene terephthalate) (PET) surface using an atmospheric pressure argon plasma source for activation [30].

In recent years, we have explored the application of low-temperature atmospheric pressure helium plasma sources to activate polymer surfaces for grafting of various monomers. For laboratory experiments with fluoropolymers such as ETFE as the substrate, the application of the handheld plasma source proved to be very efficient because surface areas of a few square centimeters could be activated in approximately 1 min by sweeping the plasma jet over the surface. Samples were reacted with ambient air to enable the formation of (hydro)peroxides and subsequently subjected to grafting. Successful grafting was indicated in ATR-IR spectra for many different vinyl monomers. Based on the determination of the surface density of peroxide species, the grafting density could be estimated to be high enough to reach the brush regime [31]. The technique was used to grow brushes of polyelectrolytes on fluoropolymers as well as on polyolefins, which enabled us to tune the surface wettability in a wide range, independent of the chemical composition of the substrate (see Chapter 4). Furthermore, grafting after plasma activation was applied to fluoropolymer films prestructured by thermal imprint of a silicon stamp in order to study the combined effects of surface topography and chemistry on anisotropic wetting properties [32].

Formation of patterns of brushes in the case of plasma activation is straightforward because simple blocking of the surface with a stencil mask or a thin foil is sufficient to inhibit the access of the plasma to the surface. Parts of the surface were covered with an adhesive polyimide (Kapton) foil, which resists the plasma treatment, in order to define 1-mm-wide and a few centimeters long stripes of high wettability on a hydrophobic fluoropolymer substrate, and possible application of this strategy in planar microfluidics was demonstrated [31]. To increase resolution, a hard mask etched into silicon was used during plasma activation to define 200-μm-sized squared fields. Figure 2.10A shows the plasma activation process. The sample was then grafted with 4VP and derivatized with iodomethane to form a charged, hydrophilic polyelectrolyte brush. When positioning the sample above a beaker with boiling water, the grafted areas became clearly visible due to the different wettability of the support polymer and

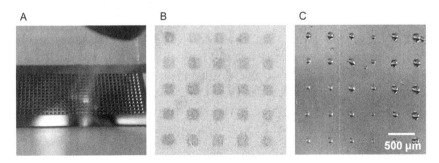

*Figure 2.10 **Micro-grafting using atmospheric pressure plasma.** (A) Activation of a fluoropolymer foil through a silicon stencil mask using an atmospheric helium plasma jet. (B) After grafting of 4VP and reaction with iodomethane to form a strong polyelectrolyte, distinct differences in surface chemistry are evident from differences in wetting in water vapor. (C) After immersion in water, droplets remain only in the grafted areas.*

the grafted areas (Figure 2.10B). After immersion in an aqueous solution for a few seconds, water droplets persisted only on the grafted areas (Figure 2.10C). Such processes are interesting for site-selective binding of active species, for instance, for application as biofunctional arrays.

2.5.2 Accelerated Ions

Depending on their energy, accelerated ions can preferentially create radicals on the surface or in the bulk of the polymer, and exposed samples may be used to achieve preferential grafting at the surface or in the bulk. For example, Ar^+ ions accelerated to relatively low energies of 300 keV in an ion implanter were used to activate surfaces of high-density polyethylene (HDPE). After exposure to air, grafting of AA was demonstrated. Shadowing with a metal mask allowed production of features in the 10-μm range as demonstrated after binding of a fluorescent dye to the grafted PAA [33]. Selective binding of biotin to similar samples was used to obtain biofunctional structures. This functionalization was verified by specific binding of labeled streptavidin [34]. In a different approach, large-area exposure to argon ions was used to create peroxide initiators for the graft polymerization of a monomer with a photosensitive side group. The grafted polymer could then be selectively activated by light to bind biotin and eventually to form biofunctional streptavidin patterns [35].

At higher energies, the penetration depth increases, leading to increased radical formation in the bulk. Similar to high-energy photon or e-beam exposures, thicker masks are needed for structuring and high aspect ratio absorbers are required to obtain microstructured samples. Proton beams with energies up to 3400 keV were used to irradiate HDPE films that were then grafted with AA or acrylonitrile [36]. Structuring in this case was achieved by placing a mask cut from a polyethylene film on the sample during exposure. Structure resolution in the range of 100 μm was demonstrated, but in order to improve resolution, the use of hard metal masks was recommended. Furthermore, multiple grafting with different monomers was demonstrated by sequential exposure and grafting steps.

Similar to electrons in e-beam writers, focused ion beams could possibly be used to write arbitrary radical patterns for subsequent graft polymerization. However, such an approach has seemingly not been reported in the literature. It is assumed that analogous to e-beam exposure, the required dose for sufficient radical generation would be comparably low, and that charging could limit the pattern resolution.

2.5.3 Swift Heavy Ions

Exposure of polymer films to accelerated heavy ions followed by etching processes is a well-known and industrially used method to obtain nanoporous membranes. Bombardment of polymer films with ions accelerated in the range of $1-10$ MeV leads primarily to bond breaking along the track of the ion. Chemical etching with a selective solvent is then used to open channels along these tracks. These so-called track-etched membranes find application in membrane technology, as materials with special optical properties and as templates to produce, for instance, metal nanowires and nanotubes by electroplating [37].

From the standpoint of nanostructuring, ion track membranes are unique in that every track can be seen as an individual nanostructure with an extreme aspect ratio: The length of the channel is usually in the 10- to 100-µm range and spans the whole thickness of the exposed film. The width of the channels, typically in the range of 100 nm to several micrometers, is defined by the ion and its energy as well as the etching parameters after ion bombardment. Furthermore, the polymer and the process parameters influence the exact shape of the channel formed [38].

In the context of controlling the functionality of the membrane, grafting of functional polymers at the inner side of the ion tracks is very attractive [39]. Three different routes can be followed, as schematically depicted in Figure 2.11.

In the simplest option, grafting can be performed directly after exposure—that is, without opening of the channels—analogous to bulk grafting of e-beam or x-ray exposed films. Grafting of styrene or methyl methacrylate was performed in PVDF films directly after exposure with different ions. Compared to grafting performed after exposure to γ-rays, higher grafting yields and more heterogeneous and intrinsically anisotropic membranes were obtained [40]. PVDF membranes modified by styrene grafting and sulfonation analogous to the process described in Section 2.4.3 showed proton conductivities similar to those of Nafion [41]. In membranes produced from ETFE films in a similar process, the proton conductivity perpendicular to the film

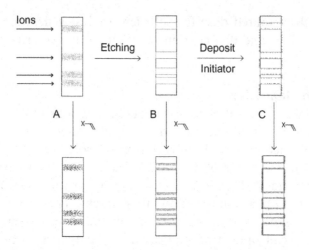

Figure 2.11 **Generation of graft membranes based on bombardment with heavy ions that leave ion tracks in a polymer film.** *(A) Radicals created in the bombardment are directly used as polymerization initiators. (B) Radicals left over after the etching step are used as initiators. (C) After etching of the ion tracks, a polymerization initiator is bound to the entire surface and used for surface graft polymerization.*

surface was three times higher than that in the direction parallel to the surface [42].

During ion bombardment, secondary electrons produced cause additional damage including radical formation in the region around the track. It was found that directly after the etching of the tracks, enough radicals were present to still start a grafting reaction (Figure 2.11B). This was demonstrated for grafting AA in PVDF [43] or PP [44], as well as GMA in PET [45]. The direct use of the radicals as initiators is advantageous because the grafting proceeds uniquely inside the pores. However, the time allowed between exposure, track etching, and grafting is limited by the lifetime of the radicals inside the exposed films.

As a third option, a preformed track-etched membrane may be functionalized by binding of initiators followed by controlled or free radical polymerization (Figure 2.11C). Such techniques are described in more detail in Chapter 3. They have been adapted to ion track-etched membranes by various groups. For instance, PHEMA and PNIPAAm were copolymerized by ATRP inside tracks etched into a PET membrane to obtain a membrane with dual environmental sensitivity [46].

2.6 CONCLUSIONS

The broad variety of radiation sources that can be used for activation of polymers generates a wealth of possibilities for producing grafted polymer structures. Surface or bulk grafting is achievable with the same base polymer and grafted monomer by careful selection of activation and reaction conditions. Selective blocking of the radiation can in all cases be used to define structures; however, the resolution is often limited when relying on shadow masks. When the radiation can be focused or interference techniques are applicable, higher resolution in the sub-micrometer to nanometer scale can be achieved.

REFERENCES

[1] Clough RL. High-energy radiation and polymers: a review of commercial processes and emerging applications. Nucl Instrum Methods Phys Res B 2001;185:8–33.

[2] Drobny JG. Ionizing Radiation and Polymers: Principles, Technology, and Applications. Amsterdam: Elsevier; 2012.

[3] Schnabel W. Polymers and Electromagnetic Radiation: Fundamentals and Practical Applications. Weinheim: Wiley–VCH; 2014.

[4] Deng Q, Chen Y, Sun W. Preparation of polymer brushes from poly(vinylidene fluoride) surfaces by UV irradiation pretreatment. Surf Rev Lett 2007;14(1):23–30.

[5] Kyomoto M, Moro T, Takatori Y, Kawaguchi H, Nakamura K, Ishihara K. Self-initiated surface grafting with poly(2-methacryloyloxyethyl phosphorylcholine) on poly(ether-ether-ketone). Biomaterials 2010;31(6):1017–24.

[6] Kyomoto M, Moro T, Yamane S, Hashimoto M, Takatori Y, Ishihara K. Poly(ether-ether-ketone) orthopedic bearing surface modified by self-initiated surface grafting of poly(2-methacryloyloxyethyl phosphorylcholine). Biomaterials 2013;34(32):7829–39.

[7] Baudin C, Renault JP, Esnouf S, Palacin S, Berthelot T. V.U.V. grafting: an efficient method for 3D bulk patterning of polymer sheets. Polym Chem 2014;5(8):2990–6.

[8] Lippert T. Interaction of photons with polymers: from surface modification to ablation. Plasma Process Polym 2005;2(7):525–46.

[9] Yamato M, Konno C, Koike S, Isoi Y, Shimizu T, Kikuchi A, et al. Nanofabrication for micropatterned cell arrays by combining electron beam-irradiated polymer grafting and localized laser ablation. J Biomed Mater Res A 2003;67A(4):1065–71.

[10] Khorasani MT, Mirzadeh H, Sammes PG. Laser surface modification of polymers to improve biocompatibility: HEMA grafted PDMS, in vitro assay—III. Radiat Phys Chem 1999;55(5-6):685–9.

[11] Auzelyte V, Dais C, Farquet P, Grutzmacher D, Heyderman LJ, Luo F, et al. Extreme ultraviolet interference lithography at the Paul Scherrer Institut. J Micro-Nanolithogr Mems Moems 2009;8:2.

[12] Langner A, Paivanranta B, Terhalle B, Ekinci Y. Fabrication of quasiperiodic nanostructures with EUV interference lithography. Nanotechnology 2012;23(10):6.

[13] Wu T, Efimenko K, Vlcek P, Subr V, Genzer J. Formation and properties of anchored polymers with a gradual variation of grafting densities on flat substrates. Macromolecules 2003;36(7):2448–53.

[14] Neuhaus S, Padeste C, Solak HH, Spencer ND. Functionalization of fluoropolymer surfaces with nanopatterned polyelectrolyte brushes. Polymer 2010;51(18):4037–43.

[15] Farquet P, Padeste C, Solak HH, Gursel SA, Scherer GG, Wokaun A. Extreme UV radiation grafting of glycidyl methacrylate nanostructures onto fluoropolymer foils by RAFT-mediated polymerization. Macromolecules 2008;41(17):6309–16.

[16] Farquet P, Kunze A, Padeste C, Solak HH, Guersel SA, Scherer GG, et al. Influence of the solvent viscosity on surface graft-polymerization reactions. Polymer 2007;48(17):4936–42.

[17] Padeste C, Farquet P, Solak HH. Surface relief polymer structures grafted onto polymer films. Microelectron Eng 2006;83(4–9):1265–8.

[18] Malek CK, Saile V. Applications of LIGA technology to precision manufacturing of high-aspect-ratio micro-components and -systems: a review. Microelectron J 2004;35(2):131–43.

[19] Chapiro A, Jendrychowska-Bonamour AM, Mizrahi S. Preparation of mosaic membranes by radiochemical grafting in polytetrafluoroethylene films. Eur Polym J 1976;12(11):773–80.

[20] Gursel SA, Padeste C, Solak HH, Scherer GG. Microstructured polymer films by X-ray lithographic exposure and grafting. Nucl Instrum Methods Phys Res B 2005;236:449–55.

[21] Farquet P, Padeste C, Borner M, Ben Youcef H, Gursel SA, Scherer GG, et al. Microstructured proton-conducting membranes by synchrotron-radiation-induced grafting. J Membr Sci 2008;325(2):658–64.

[22] Nasef MM, Hegazy ESA. Preparation and applications of ion exchange membranes by radiation-induced graft copolymerization of polar monomers onto non-polar films. Prog Polym Sci 2004;29(6):499–561.

[23] Gajos K, Padeste C, et al. Manuscript in preparation, 2015.

[24] Bogdanovitch B, Senioukov V, Koroliov A, Simonov K. Application of low energy electron beams for technology and medicine. Proceedings of the 1999 Particle Accelerator Conference (Cat. No. 99CH36366), vol. 4. 1999. p. 2570–2.

[25] Gubler L. Polymer design strategies for radiation-grafted fuel cell membranes. Adv Energy Mater 2014;4(3).

[26] Rager T. Structured radiation-grafted polymer films and membranes. J Appl Polym Sci 2006;100(1):292–4.

[27] Desmet T, Morent R, De Geyter N, Leys C, Schacht E, Dubruel P. Nonthermal plasma technology as a versatile strategy for polymeric biomaterials surface modification: a review. Biomacromolecules 2009;10(9):2351–78.

[28] Mortazavi M, Nosonovsky M. A model for diffusion-driven hydrophobic recovery in plasma-treated polymers. Appl Surf Sci 2012;258(18):6876–83.

[29] Dogue ILJ, Mermilliod N, Foerch R. Grafting of acrylic acid onto polypropylene—Comparison of two pretreatments: gamma-irradiation and argon plasma. Nucl Instrum Methods Phys Res B 1995;105(1–4):164–7.

[30] Wang JK, Liu XY, Choi HS. Graft copolymerization kinetics of acrylic acid onto the poly(ethylene terephthalate) surface by atmospheric pressure plasma inducement. J Polym Sci B Polym Phys 2008;46(15):1594–601.

[31] Neuhaus S, Padeste C, Spencer ND. Functionalization of fluropolymers and polyolefins via grafting of polyelectrolyte brushes from atmospheric-pressure plasma-activated surfaces. Plasma Process Polym 2011;8(6):512–22.

[32] Neuhaus S, Spencer ND, Padeste C. Anisotropic wetting of microstructured surfaces as a function of surface chemistry. ACS Appl Mater Interfaces 2012;4(1):123–30.

[33] Kim DK, Choi YM, Jung CH, Kwon HJ, Choi JH, Nho YC, et al. Patterned grafting of acrylic acid onto polymer substrates. Polym Adv Technol 2009;20(3):173–7.

[34] Kim DK, Ganesan R, Jung CH, Hwang IT, Choi JH, Kim JB, et al. Micropatterning of proteins on ion beam-induced poly(acrylic acid)-grafted polyethylene film. Polym Adv Technol 2011;22(12):1989−92.

[35] Yun JM, Jung CH, Kim DK, Hwang IT, Choi JH, Ganesan R, et al. Photosensitive polymer brushes grafted onto PTFE film surface for micropatterning of proteins. J Mater Chem 2010;20(10):2007−12.

[36] Taniike A, Kida Y, Furuyama Y, Kitamura A. Fabrication of a polymer with three-dimensional structure by the ion beam graft polymerization method. Nucl Instrum Methods Phys Res B 2011;269(24):3237−41.

[37] Apel P. Swift ion effects in polymers: Industrial applications. Nucl Instrum Methods Phys Res B 2003;208:11−20.

[38] Apel P. Track etching technique in membrane technology. Radiat Meas 2001;34 (1−6):559−66.

[39] Barsbay M, Guven O. Grafting in confined spaces: Functionalization of nanochannels of track-etched membranes. Radiat Phys Chem 2014;105:26−30.

[40] Betz N. Ion track grafting. Nucl Instrum Methods Phys Res B 1995;105(1−4):55−62.

[41] Clochard MC, Berthelot T, Baudin C, Betz N, Balanzat E, Gebel G, et al. Ion track grafting: a way of producing low-cost and highly proton conductive membranes for fuel cell applications. J Power Sources 2010;195(1):223−31.

[42] Kimura Y, Chen J, Asano M, Maekawa Y, Katakai R, Yoshida M. Anisotropic proton-conducting membranes prepared from swift heavy ion-beam irradiated ETFE films. Nucl Instrum Methods Phys Res B 2007;263(2):463−7.

[43] Cuscito O, Clochard MC, Esnouf S, Betz N, Lairez D. Nanoporous β-PVDF membranes with selectively functionalized pores. Nucl Instrum Methods Phys Res B 2007;265 (1):309−13.

[44] Mazzei R, Bermúdez GG, Chappa VC, del Grosso MF, Fernandez A. Grafting on nuclear tracks using the active sites that remain after the etching process. Nucl Instrum Methods Phys Res B 2006;251(1):99−103.

[45] Soto Espinoza SL, Arbeitman CR, Clochard MC, Grasselli M. Functionalization of nanochannels by radio-induced grafting polymerization on PET track-etched membranes. Radiat Phys Chem 2014;94(0):72−5.

[46] Kai P, Ruimin R, Haizhu L, Bing C. Preparation of dual stimuli-responsive PET track-etched membrane by grafting copolymer using ATRP. Polym Adv Technol 2013;24(1):22−7.

CHAPTER 3

Initiator Immobilization Strategies for Structured Brushes

To obtain micro- or nanopatterned brushes, patterning strategies for initiators must be established. Whereas radiation-induced and radiation-assisted free radical polymerization schemes were discussed in Chapter 2, the focus here is mainly on the many reports on controlled and benzophenone-mediated polymerization schemes. This chapter concentrates on initiator immobilization strategies with the possibility to obtain patterned brushes. After a general overview of initiator patterning methods, the chapter is structured according to the polymerization strategy following initiator immobilization; atom transfer radical polymerization, reversible addition fragmentation transfer polymerization, as well as benzophenone-mediated polymerization are covered.

3.1 GENERAL INITIATOR PATTERNING STRATEGIES

Four general strategies for initiator patterning are of main interest for the creation of structured polymer brushes (Figure 3.1). In lift-off-based structuring (Figure 3.1A), a thin resist is first patterned (step 1), followed by immobilization of initiators (step 2). Patterned initiator layers are then obtained after lift-off (step 3)—that is, dissolution of the resist layer, which also removes the initiator on top of the resist. To structure the resist layers, common methods such as photo-, e-beam-, and various types of nanolithography can be employed. Lift-off methods are well established on silicon wafers. However, they are difficult to implement on polymer surfaces because the adhesion of the resist is often insufficient and equipment needs to be adapted to enable lithographic exposures on polymer samples. In addition, coupling of initiators is not as straightforward as on silicon substrates.

Whereas the lift-off-based method can in principle yield patterns down to the tens of nanometers range, microcontact printing of initiators (Figure 3.1B) results in larger patterns in the micrometer range. A soft

Polymer Micro- and Nanografting. DOI: http://dx.doi.org/10.1016/B978-0-323-35322-9.00003-6

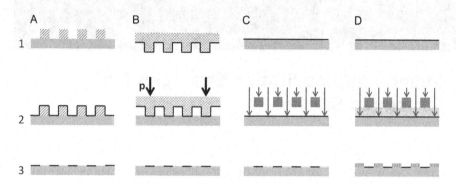

*Figure 3.1 **General initiator patterning strategies.** (A) Lift-off-based structuring; (B) micro-contact printing; (C) selective deactivation of initiators using UV light; (D) selective activation of initiators using UV light in the presence of monomer solution. See text for details.*

stamp, usually made of polydimethylsiloxane (PDMS), is coated with an initiator (step 1). The stamp is then brought into conformal contact with the surface by applying slight pressure (step 2). The initiator is thereby transferred like ink to the substrate and usually immobilized by silane coupling chemistry (step 3). The microcontact printing method is also applicable to polymer substrates because the soft stamp allows for conformal contact with slightly rough surfaces as well. Surface pretreatments, however, are required to ensure covalent immobilization of the initiator.

Using photomasks, selective, localized deactivation (Figure 3.1C) or activation (Figure 3.1D) of initiator layers is possible. The pattern resolution depends on the photomask dimensions and illumination conditions. For example, selective *deactivation* is used in atom transfer radical polymerization (ATRP) strategies as discussed later. Upon ultraviolet (UV) irradiation, the initiator loses its activity—for example, by cleavage of the halogen transfer atom. After UV patterning of the initiator, polymerization can be carried out from the substrates applying standard methods. *Activating* initiators with light, as commonly done in reversible addition fragmentation transfer (RAFT) polymerizations, requires the presence of the monomer solution already during the irradiation step. In contrast to the selective deactivation method, the polymerization in this case is carried out concurrently with the irradiation. Consequently, when applying method D (see Figure 3.1) for structured grafting, the resolution is additionally limited by the presence of the monomer solution between substrate and mask. In particular, high-resolution mask aligners and exposure techniques cannot be applied in this case.

3.2 PATTERNING STRATEGIES FOR ATOM TRANSFER RADICAL POLYMERIZATION

Most reports on ATRP from polymer surfaces deal with grafting of unpatterned polymers. However, patterning strategies using UV light reported for ATRP initiators on silicon can be generalized and applied to polymer substrates as well. Therefore, this section first introduces some examples on silicon substrates and then discusses initiator immobilization strategies on polymer substrates, bearing in mind that once immobilized, the initiators can still be patterned using UV light. This *"selective initiator deactivation"* strategy (see Figure 3.1C) is applicable to most examples of ATRP; however, this section also presents two alternative literature procedures relying on the *selective activation* of surface areas. For reference, the strategies are summarized in Figure 3.2.

3.2.1 Initiator Immobilization and Patterning on Silicon Substrates

On silicon substrates, initiators are most commonly coupled to the surface using silane chemistry. Iwata and coworkers prepared monolayers of 3-(2-bromoisobutyryl)propyl dimethylchlorosilane (BDCS) on silicon for subsequent ATRP of 2-methacryloyloxyethyl phosphorylcholine [1]. The initiator layer was patterned by irradiation with UV

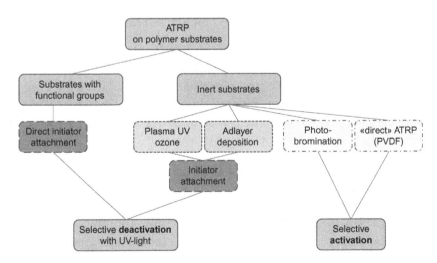

Figure 3.2 Strategies for immobilizing and patterning ATRP initiators on polymer substrates with functional groups or on inert polymer substrates.

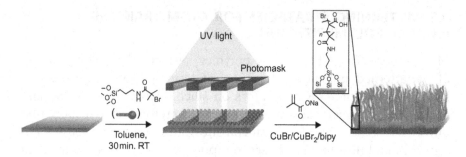

*Figure 3.3 **Immobilization of an ATRP initiator on a silicon wafer.** By irradiation through a mask using UV light, bromine is cleaved and the initiator is therefore deactivated. ATRP only takes place in regions shadowed by the photomask. Source:* Redrawn from Tugulu *et al.* [2], with permission from John Wiley & Sons Inc.

light (185 nm, 15 W) through a transmission electron microscopy (TEM) grid. In the irradiated areas, the initiator was deactivated by bromine cleavage as proven by the disappearance of the Br signal in x-ray photoelectron spectroscopy. Moreover, after polymerization, the signals from the poly(2-methacryloyloxyethyl phosphorylcholine) graft polymer were observed only in non-irradiated areas.

A similar approach was adapted to prepare poly(methacrylic acid) (PMAA) brushes on silicon using TEM grids as masks. 2-Bromo-2-dimethyl-*N*-[3-(trimethoxysilyl)propyl]propanamide was employed as initiator. Patterned PMAA brushes with a resolution in the range of 100 μm were obtained, which were then used for templating the formation of microstructured calcite films (Figure 3.3) [2].

Considering the chemical structure of ATRP initiators, selective initiator deactivation using UV light appears to be applicable to this type of initiator, independent of the type of support material. However, optimal experimental conditions for initiator deactivation on polymer surfaces and quality of the obtained pattern need to be assessed from case to case.

3.2.2 Polymers with Functional Groups for Initiator Immobilization

Initiators can be attached straightforwardly to polymers with hydroxyl or amine functionalities. A typical example for the reaction of 2-bromoisobutyryl bromide with surface functional groups in the presence of an organic base is illustrated in Figure 3.4.

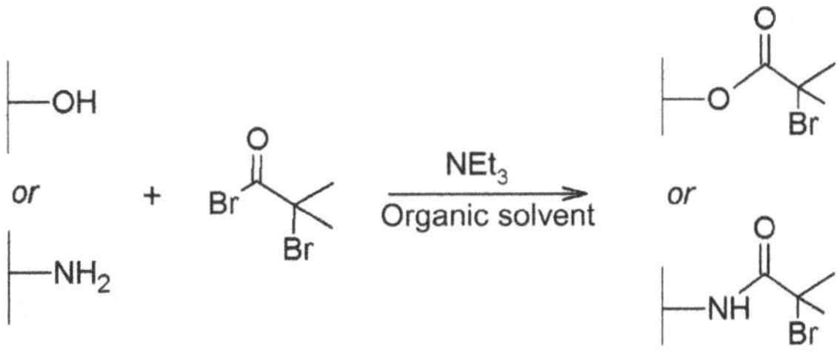

Figure 3.4 Immobilization of ATRP initiator 2-bromoisobutyryl bromide on polymer surfaces with hydroxyl or amine functionality.

This strategy was used, for instance, for grafting poly[poly(ethylene glycol) methacrylate] from regenerated cellulose [3]. Similarly, a general attempt at surface modification of polysaccharides was made by growing PMMA from chitosan films and from different types of cellulose substrates [4]. The modification of poly(ethylene terephthalate) (PET) fibers and fabrics with poly(styrene) required an additional surface modification step. PET was first reacted with 1,2-diaminoethane in methanol to provide amine and hydroxyl functional groups on the surface by aminolysis of the PET main chain. Experimental conditions were optimized to yield a sufficient number of amines (as determined by colorimetric titration) while minimizing PET chain scission and concomitant weight loss of the sample. In a second step, 2-bromoisobutyryl bromide was attached as illustrated in Figure 3.4 [5].

3.2.3 Initiator Immobilization on Inert Polymer Substrates

On inert polymer substrates, additional preparation steps are required to enable initiator immobilization. For instance, the surface of PDMS stamps was oxidized with oxygen plasma, turning the $Si\text{-}CH_3$ groups into Si-OH groups [6]. On these hydroxyl groups, a trichlorosilane ATRP initiator was attached. Aqueous ATRP of 2-(methacryloyloxy)-ethyl trimethylammonium chloride (METAC) yielded a strong cationic polyelectrolyte brush, which was used to selectively transfer perchlorate anions to flat PMETAC-covered surfaces. As a result, chloride anions were locally replaced by perchlorate anions on the flat surface, and the swellability of the polymer brush was selectively changed.

Silanol groups on PDMS could also be produced by oxidizing the surface in a UV/ozone generator [7]. The surfaces were then reacted with gaseous trichlorosilane ATRP initiator. The amount of grafted poly(acrylamide) was highest on substrates that had been treated for 15 min with UV/ozone, whereas considerable material damage became evident after a treatment duration of 30 min. On surfaces with poly(acrylamide) grafts, irreversible lysozyme adsorption was shown to be reduced by a factor of 20.

PET and poly(ethylene naphthalate) (PEN) were activated using oxygen plasma [8]. Patterns of self-assembled monolayers of trichlorosilane initiators were then deposited using microcontact printing. Microstructured thermoresponsive poly(N-isopropylacrylamide) (PNIPAM) brushes were generated by ATRP.

The strategy proposed by Yameen and coworkers [9] is more generally applicable as demonstrated by the successful grafting on five technologically important substrates: poly(propylene) (PP), poly(ether ether ketone) (PEEK), PET, poly(tetrafluoroethylene) (PTFE), and poly(4,4′-oxydiphenylene pyromellitimide) (PI). The approach involved poly(allylamine) deposition on these substrates by pulsed plasma polymerization. The amine functionalities of the resulting adlayer were then used for the attachment of ATRP initiators.

In another approach, inert isotactic polypropylene (i-PP) was activated by photobromination [10]. A Pyrex glass tube with polymer films was purged with nitrogen, filled with bromine vapor, and then irradiated with a medium-pressure mercury vapor lamp. The surface-bound bromine was used directly for the initiation of ATRP of PNIPAM. Note that patterning in this case could also be implemented using a photomask during the bromination reaction, resulting in a selective *activation* of the surface.

The use of *reverse* ATRP enlarges the scope of substrate polymers for controlled polymerizations even further because the required radical initiators can in many cases be provided more easily than ATRP initiators on surfaces of inert polymers. For example, poly(vinylidene fluoride) microfiltration membranes were irradiated with UV light and then exposed to air to create hydro(peroxide) species [11]. These were then used to initiate the reverse ATRP of methyl methacrylate in the presence of copper(I) chloride, 2,2′-bipyridine, and benzoylperoxide (Figure 3.5). Similar reaction schemes are applicable to (hydro)peroxide patterns created directly on polymer surfaces using the lithographic methods discussed in Chapter 2.

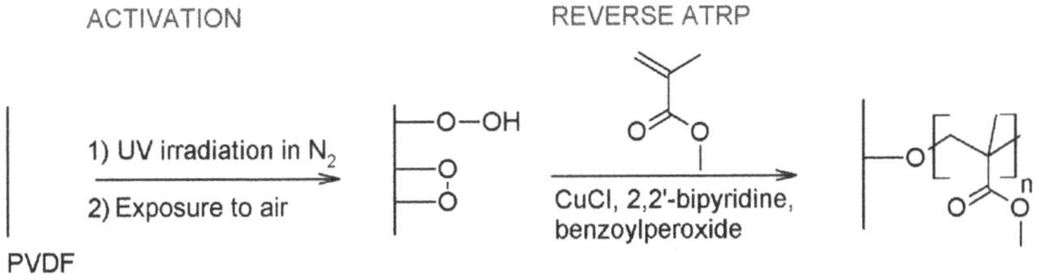

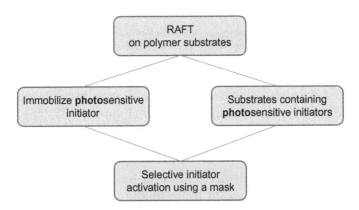

Figure 3.5 Reverse ATRP of methyl methacrylate on PVDF microfiltration membranes activated by UV irradiation [11].

3.3 PATTERNING STRATEGIES FOR REVERSIBLE ADDITION FRAGMENTATION TRANSFER POLYMERIZATION

RAFT polymerizations can be initiated using light-activatable initiators. Therefore, two main strategies are applicable to obtain patterned brushes via RAFT polymerization (Figure 3.6):

- Immobilization of a photosensitive RAFT initiator on the substrate
- Preparation of substrates containing photosensitive RAFT initiators

In the following, examples for the two approaches are provided and discussed.

Figure 3.6 Strategies for obtaining patterned brushes via RAFT polymerization from polymer substrates.

3.3.1 Site-Selective Polymerization Using a Light-Sensitive Initiator (Photoiniferter)

Nakayama and coworkers presented two different strategies for obtaining patterned polymer brushes on PET and PS [12]. In the first approach, a styrene-based polymer with photoreactive RAFT initiators was synthesized and spincoated on PET (Figure 3.7A). Alternatively,

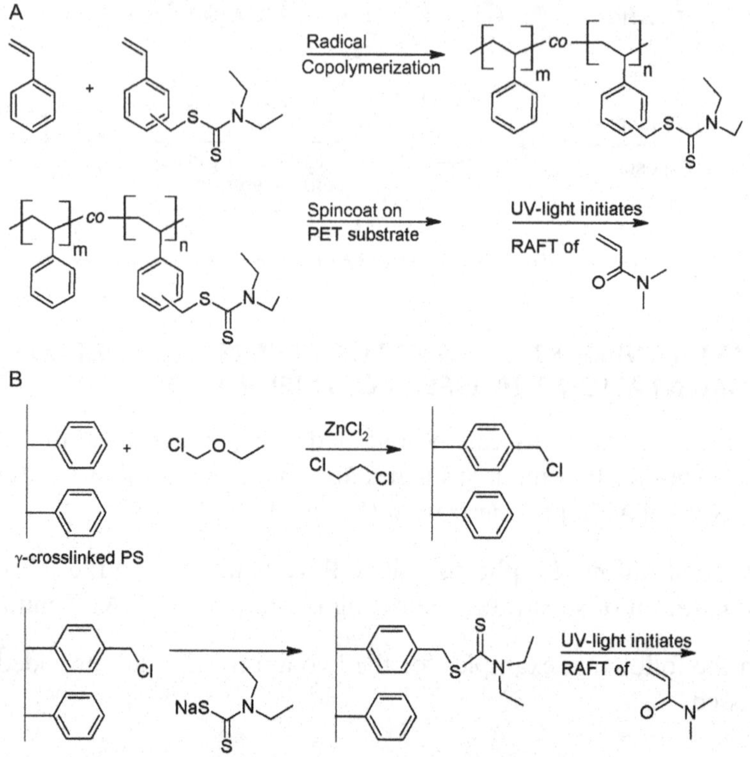

Figure 3.7 Strategies for obtaining structured graft polymers via RAFT, summarized from Nakayama and Matsuda [12]. (A) Copolymerization of styrene and vinylbenzyl N,N-diethyldithiocarbamate. Spincoating of the copolymer on a PET substrate. UV-initiated RAFT of N,N-dimethylacrylamide. (B) Chloromethylation of γ-cross-linked PS. Partial derivatization with N,N'-diethyldithiocarbamate groups. UV-initiated RAFT of N,N-dimethylacrylamide.

cross-linked PS was chloromethylated and then partially derivatized with N,N'-diethyldithiocarbamate groups (Figure 3.7B). The immobilized moieties are called "iniferters," which is the common abbreviation for a molecule acting as initiator, transfer agent, and terminator.

Surface photograft polymerizations of, for instance, N,N'-dimethylacrylamide were then initiated by irradiating the substrates in monomer solution under nitrogen atmosphere. Dimensional control of the graft polymers was obtained by selectively activating the surface using projection metal masks with 60 and 100 μm line features. Square patterns of hydrophilic graft polymers were produced for selectively seeding cells. Cross-hatched line patterns consisting of two different graft polymers were prepared in two subsequent grafting reactions. Individual staining of the graft polymers was used to prove the fidelity and reliability of this patterning method.

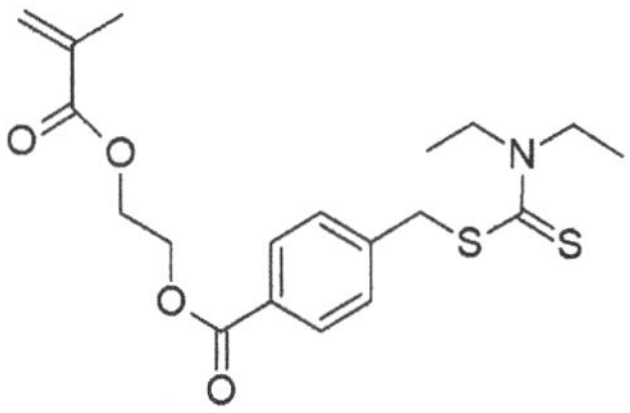

*Figure 3.8 **Methacrylated photoiniferter [13].** To be used in heat-curable monomer formulations for the preparation of iniferter-modified substrates. From the resulting materials, graft polymerizations can be initiated using UV light.*

3.3.2 Selective Polymerization Using Substrates Containing a Light-Sensitive Initiator

A methacrylated photoiniferter was synthesized for the preparation of tailor-made substrates for photografting (Figure 3.8) [13]. Substrates containing the photoiniferter were prepared by mixing different methacrylate monomers with a dimethacrylate cross-linker and benzoyl peroxide for thermal initiation of the polymerization. The solution was cast and heated to 50°C for 10 hrs to polymerize and cross-link the film. On these cast films, patterned graft copolymers with feature sizes down to 20 μm and heights in the range of several tens of micrometers were created by irradiating the substrates through a mask in the presence of monomer mixtures, thereby initiating RAFT polymerization from the immobilized iniferter. Because the graft polymer height increased almost linearly with irradiation time, thickness control was easily achieved by variation of the reaction time.

3.4 PATTERNING STRATEGIES USING BENZOPHENONE CHEMISTRY

Benzophenone (BP) is a photoinitiator that can be employed in a number of ways for the preparation of patterned graft polymers. BP can be directly used as an initiator for radical polymerization, either via immobilization of BP on the surface or by adding BP to the monomer solution (Figures 3.9A and 3.9B). Another possibility is to synthesize initiators modified with BP, which can then be immobilized on surfaces using UV light (Figure 3.9C). Yet another alternative is to add BP to the initiator solution for BP-mediated attachment of the initiator to the surface (Figure 3.9D).

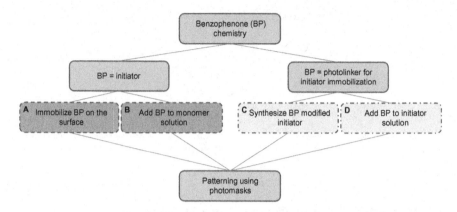

Figure 3.9 Strategies for obtaining graft polymers on polymer substrates using benzophenone (BP).

All strategies have in common that patterning is possible using photomasks.

Strategies A and B were described and compared by Stachowiak and coworkers [14]. They coined the terms "two-step sequential photografting" for strategy A and "single-step simultaneous photografting" for strategy B (Figure 3.10). Grafting was performed on macroporous polymer monoliths consisting of poly(butyl methacrylate) cross-linked by ethylene dimethacrylate.

In strategy A, BP was first immobilized on the surface of the monolith by UV irradiation. In the second step, BP was cleaved from the surface using UV light, leaving surface radicals to induce the free radical polymerization of, for instance, vinylpyrrolidone. In strategy B, monomer and BP were simultaneously added to a vial containing the substrate, and the free radical polymerization was initiated upon UV irradiation. Graft layers preventing protein adsorption to a high degree were achieved using optimized experimental parameters.

Strategy A was later applied for tailoring the surface of cyclic olefin copolymer (COC), a transparent polymer of high technological importance. Microfluidic chips made from COC were modified with hydrophilic poly[poly(ethylene glycol) methacrylate] grafts, which allowed to reduce nonspecific protein adsorption substantially [15].

Another report on the sequential photografting method deals with the grafting of poly(acrylic acid) on polypropylene membranes [16]. A linear relationship between irradiation time and average

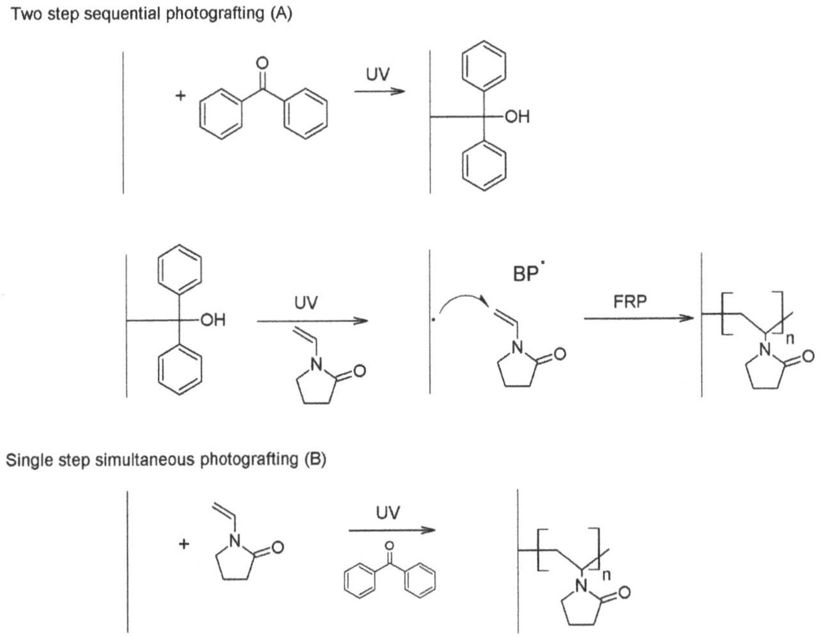

Two step sequential photografting (A)

Single step simultaneous photografting (B)

Figure 3.10 Illustration of two-step sequential photografting and single-step simultaneous photografting using ben-zophenone (strategies A and B from Figure 3.9).

length of the grafted chains was found, whereby the growth rate depended on the monomer concentration. The sequential method resulted in a fourfold greater grafting yield compared to the simultaneous method.

Strategy C (Figure 3.9) was implemented for the preparation of polypropylene with an antibacterial coating [17]. One of the most common ATRP initiators, 2-bromoisobutyryl bromide, was reacted with 4-hydroxybenzophenone in the presence of triethylamine (Figure 3.11). The BP moiety was then used as a photolinker to covalently attach the initiator to the polypropylene surface after spin coating. UV irradiation (365 nm) resulted in the attachment of the photolinker to the polypropylene backbone via hydrogen abstraction. ATRP of 2-(dimethyl-amino)ethyl methacrylate followed by quaternization allowed for the preparation of surfaces with almost 100% killing efficiency for *Escherichia coli*. A prerequisite for this high efficiency was the preparation of graft polymers with molecular weights of at least 9800 g/mol to ensure the availability of a sufficiently high surface concentration of quaternary ammonium groups.

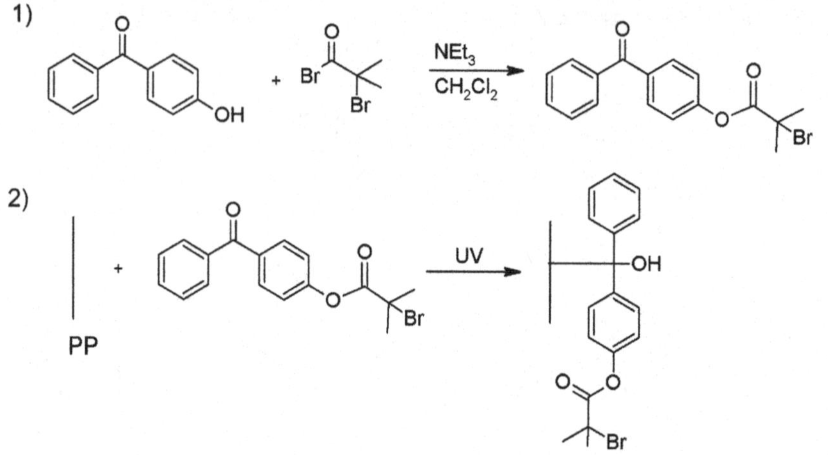

1)

2)

Figure 3.11 **Example for strategy C (Figure 3.9) as applied by Huang et al. [17].** *(1) Synthesis of a BP-modified ATRP initiator. (2) Upon UV irradiation: covalent immobilization of the ATRP initiator on poly[propylene] substrates using the BP photolinker.*

Strategy D (Figure 3.9) was used for immobilizing ATRP initiators on PE and nylon surfaces [18]. To this end, nylon or PE foils were immersed into mixtures of 4-vinylbenzyl chloride and BP and then irradiated with UV light (350 nm) for 1−3 min. Thereby, oligo(4-vinylbenzyl chloride) was grafted to the surface and later used as initiator, whereas poly(4-vinylbenzyl chloride) homopolymer was formed in solution (Figure 3.12). A significant drawback of this method is the need for extensive rinsing of the substrates and extraction of loosely bound homopolymer before the substrates can be used for ATRP.

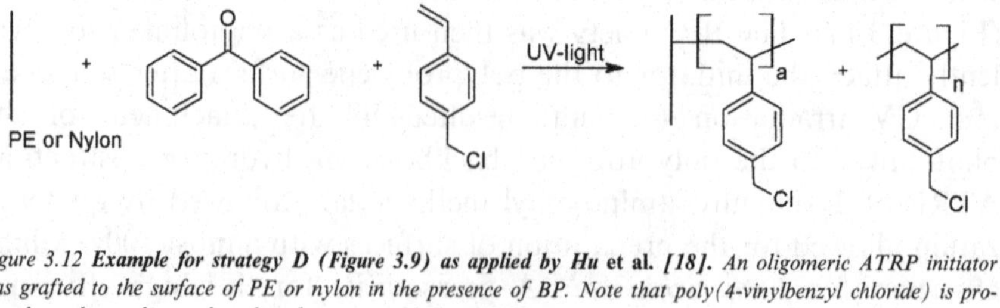

Figure 3.12 **Example for strategy D (Figure 3.9) as applied by Hu et al. [18].** *An oligomeric ATRP initiator was grafted to the surface of PE or nylon in the presence of BP. Note that poly(4-vinylbenzyl chloride) is produced as a by-product and needs to be extracted before ATRP.*

3.5 CONCLUSIONS

Various initiator strategies for growing polymer brushes rely on the use of (UV) light for immobilization, activation, or deactivation. Combination with lithographic exposure tools to obtain patterns appears to be straightforward in many cases. However, standard photolithographic equipment is often not compatible with polymeric substrates and with the chemical environment used for the involved reactions. Development of specialized equipment is therefore required to fully exploit patterned grafting techniques on polymer surfaces.

REFERENCES

[1] Iwata R, Suk-In P, Hoven VP, Takahara A, Akiyoshi K, Iwasaki Y. Control of nanobiointerfaces generated from well-defined biomimetic polymer brushes for protein and cell manipulations. Biomacromolecules 2004;5(6):2308–14.

[2] Tugulu S, Harms M, Fricke M, Volkmer D, Klok H-A. Polymer brushes as ionotropic matrices for the directed fabrication of microstructured calcite thin films. Angew Chem Int Ed 2006;45(44):7458–61.

[3] Singh N, Chen Z, Tomer N, Wickramasinghe SR, Soice N, Husson SM. Modification of regenerated cellulose ultrafiltration membranes by surface-initiated atom transfer radical polymerization. J Membr Sci 2008;311(1–2):225–34.

[4] Lindqvist J, Malmstrom E. Surface modification of natural substrates by atom transfer radical polymerization. J Appl Polym Sci 2006;100(5):4155–62.

[5] Bech L, Elzein T, Meylheuc T, Ponche A, Brogly M, Lepoittevin B, et al. Atom transfer radical polymerization of styrene from different poly(ethylene terephthalate) surfaces: films, fibers and fabrics. Eur Polym J 2009;45(1):246–55.

[6] Azzaroni O, Brown AA, Huck WTS. UCST wetting transitions of polyzwitterionic brushes driven by self-association. Angew Chem Int Ed 2006;45(11):1770–4.

[7] Xiao DQ, Zhang H, Wirth M. Chemical modification of the surface of poly(dimethylsiloxane) by atom-transfer radical polymerization of acrylamide. Langmuir 2002;18(25):9971–6.

[8] Farhan T, Huck WTS. Synthesis of patterned polymer brushes from flexible polymeric films. Eur Polym J 2004;40(8):1599–604.

[9] Yameen B, Khan HU, Knoll W, Foerch R, Jonas U. Surface initiated polymerization on pulsed plasma deposited polyallylamine: a polymer substrate-independent strategy to soft surfaces with polymer brushes. Macromol Rapid Commun 2011;32(21):1735–40.

[10] Desai SM, Solanky SS, Mandale AB, Rathore K, Singh RP. Controlled grafting of N-isopropylacrylamide brushes onto self-standing isotactic polypropylene thin films: surface initiated atom transfer radical polymerization. Polymer 2003;44(25):7645–9.

[11] Chen Y, Deng Q, Mao J, Nie H, Wu L, Zhou W, et al. Controlled grafting from poly(vinylidene fluoride) microfiltration membranes via reverse atom transfer radical polymerization and antifouling properties. Polymer 2007;48(26):7604–13.

[12] Nakayama Y, Matsuda T. Surface macromolecular architectural designs using photo-graft copolymerization based on photochemistry of benzyl N,N-diethyldithiocarbamate. Macromolecules 1996;29(27):8622–30.

[13] Luo N, Metters AT, Hutchison JB, Bowman CN, Anseth KS. A methacrylated photoinifer-ter as a chemical basis for microlithography: micropatterning based on photografting poly-merization. Macromolecules 2003;36(18):6739−45.

[14] Stachowiak TB, Svec F, Frechet JMJ. Patternable protein resistant surfaces for multifunc-tional microfluidic devices via surface hydrophilization of porous polymer monoliths using photografting. Chem Mater 2006;18(25):5950−7.

[15] Stachowiak TB, Mair DA, Holden TG, Lee LJ, Svec F, Frechet JMJ. Hydrophilic surface modification of cyclic olefin copolymer microfluidic chips using sequential photografting. J Separation Sci 2007;30(7):1088−93.

[16] Ma HM, Davis RH, Bowman CN. A novel sequential photoinduced living graft polymeriza-tion. Macromolecules 2000;33(2):331−5.

[17] Huang J, Murata H, Koepsel RR, Russell AJ, Matyjaszewski K. Antibacterial polypropylene via surface-initiated atom transfer radical polymerization. Biomacromolecules 2007;8(5):1396−9.

[18] Hu Y, Li JS, Yang WT, Xu FJ. Functionalized polymer film surfaces via surface-initiated atom transfer radical polymerization. Thin Solid Films 2013;534:325−33.

CHAPTER 4

Functional Polymer-on-Polymer Structures

In this chapter, routes toward functional polymer-on-polymer structures and their properties are described. The first section focuses on the
chemical functionality of polymer structures. Selection of the appropri-
ate chemical functionality is of significance in the design of surface
properties such as wettability and is crucial if subsequent covalent
immobilization of molecules, ranging from organic functional moieties
to DNA or proteins, is targeted. Strategies for creating polymer layers
containing amines or carboxylic acids, for example, are highlighted,
with emphasis on polyelectrolyte brushes. In the second part of this
chapter, polymer structures responsive to external stimuli such as temperature, pH, magnetic fields, light, and counterion exchange are discussed. Finally, biofunctional polymer structures with antifouling
properties or with embedded or attached biomolecules are presented.

4.1 GRAFTING FUNCTIONAL MONOMERS VERSUS POST-POLYMERIZATION MODIFICATION

Creating grafted polymer layers with a variety of chemical functional
groups is of great interest for the following reasons:

- Brushes with *carboxylic acid or amine* moieties can be used for the
  covalent attachment of molecules, particles, or proteins using well-
  established peptide coupling agents.
- *Weak polyelectrolytes* are switchable between their neutral and
  charged states by changing the pH value of the surrounding
  medium. This change is mostly accompanied by a strong change in
  wettability.
- *Strong polyelectrolytes* are of interest to create permanently charged
  surfaces. Sometimes strikingly different surface properties can be
  obtained on one and the same brush by counterion exchange.
- *Very hydrophilic brushes that swell strongly in water* are considered for
  a number of applications such as antifouling coatings or for mimicking
  natural, fluid-rich structures such as the extracellular matrix.

Polymer Micro- and Nanografting. DOI: http://dx.doi.org/10.1016/B978-0-323-35322-9.00004-8

Consequently, we have invested considerable effort in establishing a "library of functional polymer brushes." The examples discussed here were all grafted from poly(ethylene-*alt*-tetrafluoroethylene) (ETFE), poly(propylene) (PP), and poly(ethylene) (PE) using activation with extreme ultraviolet (EUV) interference lithography [1] for nanopatterned brushes or atmospheric pressure helium plasma for activating larger areas (e.g., in the context of wettability studies) [2]. The graft polymers were obtained by free radical polymerization (FRP) of commercially available monomers. Note that the strategies presented here may in some cases not be applicable in controlled radical polymerization schemes. For instance, methacrylic acid cannot be directly grafted using atom transfer radical polymerization because the polymerization catalyst can be complexed by carboxylic acid moieties.

Two main strategies were followed: *direct grafting* of monomers with the functionality of interest or grafting of a polymer brush that is then further converted via *post-modification*. For examples of both strategies, see Figure 4.1. Brushes obtained through the direct grafting strategy include weak polyelectrolytes such as poly(4-vinylpyridine) (P4VP) or poly(methacrylic acid) (PMAA) (Figures 4.1A and 4.1B). Direct grafting of monomers with the desired functionality is clearly a time- and effort-saving way of producing functional brushes. A wealth of monomers are commercially available for successful completion of this one-step procedure. However, some charged monomers are difficult to graft from hydrophobic substrates [1,3]. Grafting of diallyldimethylammonium chloride and sodium styrene sulfonate from ETFE was not successful in our experiments possibly because of the hydration sphere around the side groups of the charged monomers, which prevented the approach of the monomers to the surface. For the same reason, brushes of primary and secondary amines cannot be obtained in a straightforward manner using charged monomers such as 2-aminoethyl methacrylate hydrochloride. Neutralization of this particular monomer by adapting the pH of the solution is not an option because the molecule undergoes rearrangements at higher pH.

The limitations imposed by the incompatibility of certain monomer/substrate combinations or the lack of a suitable monomer can be overcome by the post-polymerization-modification strategy, which serves to enlarge the range of obtainable functionalities. Furthermore, modifications of the grafted polymer are possible with reagents that cannot

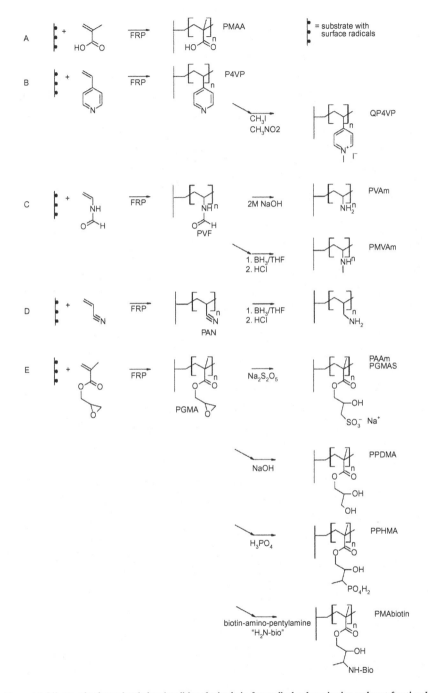

*Figure 4.1 **Library of polymer brush functionalities obtained via free radical polymerization and post-functionalization, illustrating the vast potential for creating chemical functionalities of interest.** See text for details. Source:* Figure compiled from Neuhaus *et al.* [1,2] and Padeste *et al.* [4].

be used for monomer modification because of their reactivity toward double bonds. Purification of the modified surface-bound polymer layers is comparatively simple because by-products can be washed out quite easily. As a drawback, full conversion is mostly difficult to achieve. (More information on monitoring the progress of the reaction and the detection of contaminants with attenuated total reflectance infrared spectroscopy (ATR-IR) is provided in Chapter 5). Special consideration must be given to the chemical resistance of the substrate when planning a post-functionalization step. Fluoropolymers have excellent chemical stability and are therefore not prone to degradation upon treatment with harsh chemicals, but other substrates are more susceptible to damage or chemical changes. The linkage of the polymer brush to the substrate can also be critical, even though most grafting strategies on polymer substrates lead to robust covalent attachment of the brush. Finally, the penetration of reagents into the base polymer should not be neglected, particularly if the base polymer is a reasonably good solvent for the reagents.

As shown in Figure 4.1C, primary amines can be derived from poly(vinylformamide) (PVF) brushes by basic hydrolysis. Primary or secondary amines are obtained by reduction of PVF or poly(acrylonitrile) (PAN) with borane/tetrahydrofuran (Figures 4.1C and 4.1D). Strong, quaternary amine-bearing polymer brushes are produced by quaternization of P4VP with methyl iodide (Figure 4.1B). Poly(glycidyl methacrylate) (PGMA) can be modified by a number of routes to yield brushes with different characteristics (Figure 4.1E). Treatment with sodium metabisulfite results in a strong polyelectrolyte brush bearing sulfonate groups, whereas treatment with phosphoric acid yields a weak polyelectrolyte that is amenable for three-stage pH-induced switching. Hydrolysis of PGMA with sodium hydroxide results in a polymer brush containing diol functionalities. Biofunctionality can be introduced in PGMA, for instance, in a single step by nucleophilic attack of pentylamine-functionalized biotin.

The impact of the chemical functionality of grafted polymer brushes on the wettability of the substrate was investigated for a number of neutral polymers as well as for weak and strong polyelectrolytes (Figure 4.2) [2]. It is evident that the wettability of the hydrophobic ETFE substrate can be tailored over a wide range with polymer brushes. As a general trend, surfaces with more hydrophilic brushes

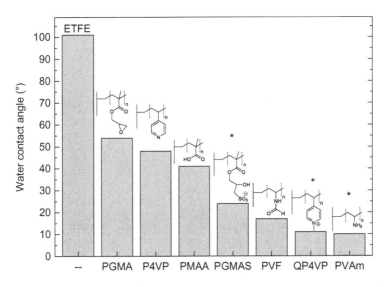

Figure 4.2 **Water contact angle as a function of chemical functionality on brush-covered surfaces.** *Brushes were grafted from ETFE via free radical polymerization. Brushes marked with an asterisk were obtained via post-functionalization. Source:* Adapted from Neuhaus *et al.* [2], with permission from John Wiley & Sons Inc.

tend to have lower water contact angles. Particularly striking is the comparison between the water-insoluble PGMA and P4VP brushes and their soluble strong polyelectrolyte counterparts (PGMAS and quaternized P4VP (QP4VP)), where the introduction of ionic groups resulted in a significant decrease in water contact angle. The lowest water contact angles ($\sim 10°$) were obtained for QP4VP and poly(vinyl-amine) (PVAm) brushes, both of which are accessible only via post-modification.

4.2 RESPONSIVE STRUCTURES

4.2.1 Responsiveness to Changes in pH

Studies of pH responsiveness are conducted mostly on unstructured polymer brushes because the primary observable response to pH changes lies in altered wettability, which can best be assessed macroscopically. However, exploiting the concomitant swelling (charged state) and partial collapse (neutral state) [5] in micro- and nanostructured brushes is of great future interest. One can, for instance, envision triggering the site-selective collection and release of particles or molecules of opposite charge on patterned polyelectrolyte brushes by changing the pH in the solution in contact with the surface.

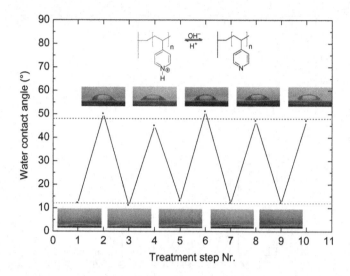

Figure 4.3 Reversible switching of the wettability of a P4VP brush-modified ETFE surface. The sample was immersed in 0.1 M HCl or H_2O prior to contact angle determination, and the process was repeated several times to demonstrate the reversibility of the change in contact angle. The resulting water CAs differed by 36°. *Source*: Adapted from Neuhaus *et al.* [2], with permission from John Wiley & Sons Inc.

We have demonstrated the reversible switching of surface wettability using the examples of P4VP and PMAA brushes [2]. In the charged form (i.e., at low pH), P4VP brushes were very hydrophilic with a water contact angle of 12°. In the neutral form, the contact angle increased to 48°, yielding a difference of 36° (Figure 4.3). The contact angle difference obtained between neutral and negatively charged PMAA surfaces was lower (21°) because PMAA remains water soluble in the neutral state. In contrast, neutral P4VP is completely insoluble and therefore expected to collapse at high pH.

Wettability switching has been more widely explored on nonpolymeric substrates, but many findings may also be applicable to polymer brushes on polymeric substrates. Patterned microstructured brushes on gold have been created by microcontact printing of self-assembled monolayers containing initiators for ATRP. After the grafting step, the resulting poly[2-(methacryloyloxy)ethyl phosphate] brushes exhibited three-stage wettability switching from 65° at low pH (fully protonated) to 13° at high pH (fully deprotonated) (Figure 4.4) [6]. However, it was reported that the direct polymerization of the phosphate-bearing monomer was difficult and prone to side reactions. We therefore suggest using the strategy outlined in Figure 4.1E to obtain phosphate-bearing polymers on polymer substrates.

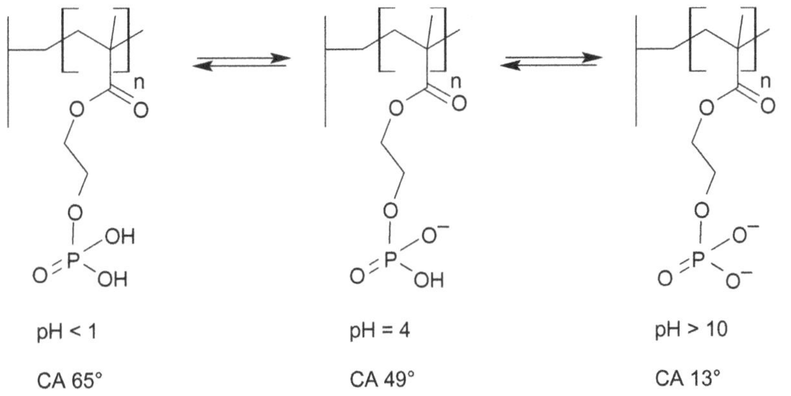

Figure 4.4 Three-stage switching of poly[2-(methacryloyloxy)ethylphosphate] as reported for brushes grown by ATRP from gold substrates [6].

4.2.2 Responsiveness to Temperature Changes

Poly(N-isopropylacrylamide) (PNIPAAm) is by far the most prominent example of a thermally responsive polymer. It undergoes a phase transition at the lower critical solution temperature (LCST), resulting in a strong decrease in hydration of the polymer. For polymer brushes, this behavior is reflected in a collapse of the structure above the LCST. Because of the proximity of the LCST (32°C) to the body temperature, PNIPAAm is considered an interesting candidate for drug release systems.

Heinz *et al.* have prepared PNIPAAm layers in a multistep procedure. Poly(ethylene oxide) (PEO) layers were deposited on silicon wafers by plasma polymerization. After plasma discharge activation of PEO, NIPAAm was grafted by free radical polymerization [7]. The brush was not patterned, but in principle masking strategies during plasma activation would be applicable. When increasing the temperature from 15° to 45°C, a smooth increase of water contact angles was observed, and a plateau was reached at 40°C. From the deflection point of the curve, an LCST of 28°C was derived for the investigated system (Figure 4.5). The change in wettability was accompanied by a change in the viscoelastic properties of the surface, as determined using a quartz crystal microbalance. Moreover, switching from a state of protein repulsion below the LCST to a state of protein adsorption above the LCST was observed. When returning to lower temperatures, proteins were desorbed again.

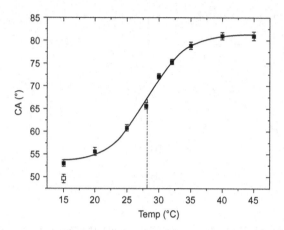

Figure 4.5 ***Advancing water contact angle as a function of temperature on a PEO surface grafted with PNIPAAm*** *[7].* *Water contact angles steadily increase with temperature; from the deflection of the curve, a LCST of 28° C can be derived. After cooling, a contact angle of about 50° was obtained (open symbol).* Source: Reproduced from Heinz *et al.* [7], with permission from ACS.

Nanopatterned brushes of PNIPAAm were prepared by our group by activating ETFE samples with EUV interference lithography followed by FRP of NIPAAm [8]. The phase transition behavior was visualized in water on dot patterns with 707-nm period using the phase signal in atomic force microscopy. The phase signal is influenced by a number of factors, all of which contribute to the energy dissipation involved in the contact between the tip and the sample [9]. A gradual transition in image contrast and signal range was observed with increasing temperature, indicating a transition from an extended to a collapsed state of the brush—that is, from a state of strong energy dissipation to a state of little energy dissipation (Figure 4.6).

4.2.3 Magneto-responsiveness

The contactless actuation of surfaces is of particular interest for remotely controlling a system. Polymer brushes with embedded magnetic nanoparticles are well suited for this purpose because the brush dimensions can be changed in a good solvent by applying an external magnetic field. Despite the attractiveness of magnetic particle-loaded systems, only a limited number of studies have been published on this subject. We present an example of structured brushes on silicon, followed by a discussion of our own efforts on this topic.

Huck's group studied the creation of magnetic iron oxide particles in poly(3-sulfopropyl methacrylate) brushes grown via ATRP on

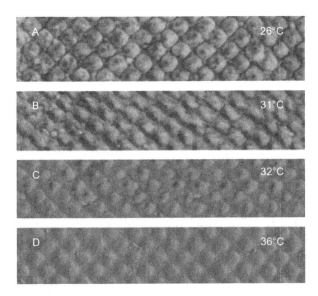

Figure 4.6 AFM phase contrast images of periodic PNIPAAm structures (dot spacing, 707 nm) created by FRP of NIPAAm from ETFE substrates activated with EUV interference lithography. The decreasing contrast with increasing temperature serves to illustrate the phase transition of PNIPAAm at approximately 32°C. Source: Reproduced from Farquet et al. [8], with permission from Elsevier Ltd.

gold-coated silicon with the initiator patterned by microcontact printing [10]. The high affinity of sulfonate groups for iron cations was then used to load the brushes with Fe^{2+} and Fe^{3+} in exchange for K^+ counterions. After slow oxidation at room temperature, the presence of iron oxides of different stoichiometries was confirmed by x-ray photoelectron spectroscopy and x-ray diffraction. The size of the particles was determined to be between 3 and 8 nm by transmission electron microscopy (TEM). The particle-loaded brush structures exhibited strong changes in both width and height upon application of a magnetic field.

In our group, polymers containing functional moieties (carboxylate and pyridine) capable of complexing iron cations were grafted from ETFE [11]. Brushes were then loaded with Fe^{2+} in 0.1 M solutions of $FeCl_2$ in water or ethanol, respectively. Finally, iron oxide nanoparticles were created by oxidation in sodium hydroxide solution (Figure 4.7). The complexation and oxidation step as well as the magnetic properties were studied in detail with attenuated total reflectance infrared spectroscopy (ATR-IR) and scanning transmission x-ray microscopy (STXM) (for analysis strategies, see Chapter 5).

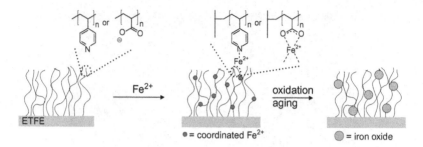

Figure 4.7 **In situ** *creation of iron oxide nanoparticles in a polymer brush matrix. Following grafting, Fe²⁺ was com-plexed by the polymer brush matrix. Iron oxide nanoparticles were created via alkaline hydrolysis and oxidation.*

Straightforward information on the brush/nanoparticle composites was obtained from TEM images and UV/vis spectroscopy of the foils grafted with polymer brushes and loaded with iron oxide (Figure 4.8). Imaging and energy dispersive x-ray analysis (EDX) confirmed the presence of tiny (~1 nm) iron-containing particles. Efforts to create larger particles by repeating the number of loading and complexation steps did not result in increased particle size. However, a strongly decreased transmittance of the foils between 300 and 600 nm was observed, indicating that the density of iron oxide particles had signifi-cantly increased (Figure 4.8B). Similar observations were made in polymer gels and polyelectrolyte multilayers [12,13]. It appears that particle nucleation is strongly favored over particle growth in poly-meric matrices.

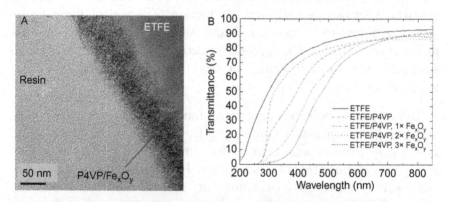

Figure 4.8 **Iron oxide nanoparticle formation in a P4VP brush matrix.** *(A) TEM image of a microtome cut of approximately 100nm thickness of a P4VP/Fe$_x$O$_y$ nanocomposite, showing an iron oxide particle-loaded brush after three treatment sequences consisting of iron complexation and oxidation. (B) UV spectra of ETFE samples modified with P4VP brushes and P4VP brush/Fe$_x$O$_y$ nanocomposites. The transmittance between 300 and 600 nm decreased strongly with the number of loading steps.*

The nanoparticles in these P4VP brushes could not be magnetized at room temperature, probably due to their small size. Because the surface-to-volume ratio is very high, surface effects are expected to play an important role. This could lead to poorly controlled, low crystallinity and poor stoichiometry of the particles. In the poly(acrylic acid) matrix, infrared and x-ray absorption spectroscopy (XAS) results strongly suggest that the composite mainly consisted of Fe^{2+} complexed by carboxylate groups rather than iron oxide particles. Considering all the findings discussed here (TEM, EDX, and UV/vis) and in Chapter 5 (ATR-IR, STXM, and XAS), it appears that the *in situ* preparation of nanoparticles in a polymer brush is a very delicate task. The confinement provided by the brush matrix can strongly limit particle growth as observed in the P4VP matrix. On the other hand, the complexes formed with the brush may be too stable and, thus, prevent particle growth (PAA matrix).

4.2.4 Responsiveness Based on Counterion Exchange

Counterion exchange is another interesting strategy for inducing changes in polyelectrolyte brush-modified surfaces. By exposing the surface to a solution containing an abundance of alternative counterions, the surface-bound counterions can be replaced. Huck and coworkers extensively studied the strong cationic polyelectrolyte poly[[3-(methacryloyloxy)ethyl]trimethyl ammonium chloride] on silicon substrates [14,15]. Different classes of counterions were identified and studied in terms of changes in nanomechanical properties and wettability:

- Counterions leading to a reversible collapse of the brush by screening of charges (e.g., Cl^-)
- Counterions leading to an irreversible collapse by strong ion pairing interactions (e.g., SO_4^{2-})
- Counterions that are not able to replace the ions in the brush because of sterical hindrance (e.g., benzyltributyl ammonium chloride)
- Counterions with strong hydrophobic (e.g., bis(trifluoromethane)sulfonimide) or hydrophilic (e.g., polyphosphate) character. Their mutual exchange provoked very pronounced changes in water contact angle from 90° to 15°.

We have switched the wettability of poly(N-methylvinylpyridinium) brushes on ETFE from approximately 15° to greater than 100° by

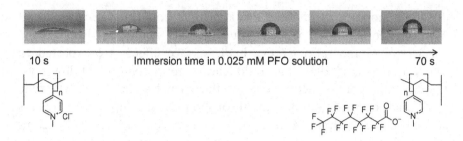

Figure 4.9 Poly(N-methylvinylpyridinium) brush on ETFE. Exchange of chloride or iodide counterions by per-fluorooctanoate counterions. The water contact angle increases with immersion time as the counterion exchange proceeds. A steady state is reached after approximately 50s immersion in PFO solution.

exchanging chloride or iodide for strongly hydrophobic perfluoroocta-noate (PFO) counterions. The time frame of the counterion exchange (from Cl^- or I^- to PFO) is in the range of minutes, as illustrated in Figure 4.9. Noticeably, the exchange was much slower when replacing PFO by chloride. These exchange processes were adapted to reversibly switch gradients in surface wettability from a hydrophilic to a hydro-phobic state [16].

4.2.5 Light Responsiveness

Adding to the range of available stimuli for influencing surface proper-ties, much effort has been devoted to photoactive molecules because light irradiation can be used as noncontact external stimulus. This stimulus can be applied locally but remotely with a precise control of light fluxes. The photochromic properties of spiropyran derivatives have probably been most heavily investigated. Joseph et al. grafted copolymers of acrylated spirobenzoypyran and N-isopropylacrylamide from immobilized benzophenone initiators on cyclic olefin copolymer (COC) [17]. The UV-induced free radical polymerization was carried out in the presence of a cross-linker (N,N'-methylenebisacrylamide). The water contact angle increased upon light irradiation, and it was speculated that a synergistic effect between the photoisomerization of the spirobenzopyran moieties and dehydration of PNIPAAm similar to the effect observed upon heating, was responsible for the change in wettability. Grafts were also prepared on micro- and nanostructured COC substrates [17]. Microstructured substrates were obtained by sol-vent vapor-induced stress crazings, whereas surface nanorods were pre-pared by using anodic aluminum oxide in an intricate templating procedure. The reversible wettability changes were found to strongly increase from flat to microstructured to nanostructured surfaces due

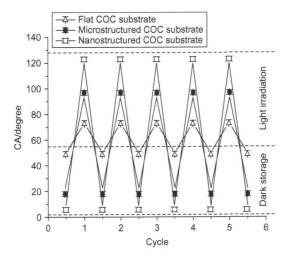

*Figure 4.10 **Reversible wettability switching upon alternating irradiation with light and storage in the dark.** The effect is weakest on flat COC, more pronounced on microstructured COC, and strongest on COC with nanorods.* Source: Redrawn from Joseph *et al.* [17], with permission from RSC Publishing.

to increased surface roughness and pinning effects (Figure 4.10). Preparation of structured coatings was not demonstrated, but it is in principle possible using masks during initiator immobilization or during the free radical polymerization.

Recently, we explored strategies for binding spiropyran moieties to structured brushes grafted from ETFE and Teflon (PTFE) surfaces in order to obtain light-sensitive structured polymer surfaces [18]. We focused on post-polymerization modification of grafted brush structures because this strategy increases the flexibility with respect to the optimization of the concentration of spiropyran moieties in the brushes. Moreover, the tedious synthesis and purification steps of spiropyran—monomer conjugates are circumvented.

A spiropyran derivative with an amine linker was synthesized and coupled to grafted brushes of PGMA and of PMAA activated with perfluorophenyl trifluoroacetate. Exposure of the modified brushes to UV light, which converts the attached moieties from the uncharged transparent spiropyran to the zwitterionic merocyanine form, rendered the brushes deep purple in color, fluorescent, and hydrophilic. The properties switched back by thermal or visible light-induced relaxation. The kinetics of the switching was found to be dependent on the chemical environment provided by the polymer brushes as well as on the polarity of added solvent.

4.3 BIOFUNCTIONAL STRUCTURES

Numerous preparation strategies of varying complexity for biofunctional coatings have been reported in the literature. The obtained coatings exhibit functions ranging from chemical binding sites for biomolecule attachment to antibiofouling properties. A selection of approaches for creating biofunctional structures on polymer substrates is presented here.

Yun *et al.* presented a method to selectively immobilize biotin on PTFE [19]. The substrate was first irradiated with argon ions, and peroxides were generated upon exposure to air. A diazoketo-functionalized methacrylate monomer was synthesized and then grafted from the activated substrate by free radical polymerization. UV exposure of the grafted layer was performed through a mask, leading to local transformations of the diazoketo groups into carboxylic acids by Wolff rearrangement. Surface chemical patterns with approximately 50-μm feature size were thus prepared, and amine-functionalized biotin was selectively immobilized on the irradiated regions. Subsequently, binding of a streptavidin conjugate was demonstrated.

In another approach, poly(propylene) plates were modified with a custom-synthesized bifunctional molecule having a photolinker and an initiator for ATRP [20]. This benzophenonyl bromoisobutyrate was attached to the substrate by irradiation with UV light (365 nm) for 30 min. After washing, N,N'-dimethylacrylamide or poly(ethylene glycol)methacrylate (PEGMA) were polymerized with ATRP. Patterning was not demonstrated, but it could be implemented by using a mask during the initiator immobilization step. Both graft polymers led to a hydrophilization of the poly(propylene) plates. In addition, protein repellent properties were observed in the case of poly[PEGMA] brushes.

Although the previously discussed methods are applicable specifically to polymer substrates, there are also strategies for surface functionalization with polymer brushes that work on a broad range of substrates. A general method that allows producing polymer brushes on silicon, gold, perfluorinated poly(ethylene-co-propylene), and poly(styrene) has been recently introduced [21]. The first step of the modification sequence was not sensitive to the substrate used. An allylamine plasma polymer was deposited by radio frequency glow discharge. Then a macroinitiator was attached to this layer by peptide coupling, followed by ATRP of N,N'-dimethylacrylamide. The poly(N,N'-dimethylacrylamide) brushes exhibited significantly reduced protein adsorption—a property that is particularly useful for poly(styrene) microtiter plates, one of the substrates used in this study.

A very interesting strategy for grafting of biofunctional polymer brushes on polymer substrates was presented by Kyomoto et al. [22]. Poly(ether ether ketone) (PEEK) and, more generally, polyaromatic ketones have excellent stability against chemicals and radiation damage and enhanced mechanical properties compared to other polymer materials. PEEK is therefore used as a bearing material in implants for orthopedic and spinal surgeries. However, surface modifications of PEEK are often necessary to decrease fouling and to prevent undesired biofilm formation. Zwitterionic poly(2-methacryloyloxyethyl phosphorylcholine) (PMPC) is suitable for this purpose because it is known for its excellent biocompatibility, as well as for antifouling and low friction properties. This material was grafted from PEEK using a technique that was coined by the authors as "self-initiated surface graft polymerization" (see Chapter 2, Section 2.3.1). This strategy uses the fact that the diphenylketone moiety in the PEEK backbone can act as a photoinitiator similar to benzophenone. The experimental procedure is straightforward and only involves immersing clean PEEK in an aqueous, degassed solution of MPC at 60°C. The grafting reaction is triggered by shining UV light (350 nm) on the flask for a given period of time. On the PMPC-modified PEEK, three individual improvements of surface properties were clearly demonstrated as a function of irradiation time (Figure 4.11). First, the water contact angle was drastically reduced from greater than 90° to approximately 10°. Second, a concomitant decrease in the coefficient of dynamic friction from 0.2 to less than 0.01 was observed, and this was due to the lubricating properties of PMPC. Third, a strong decrease in nonspecific adsorption of bovine serum albumin (BSA) was also achieved, indicating antifouling properties. The change in surface properties exhibited pronounced dependence on the grafting time, reaching an optimum steady state after reaction times of 90 min. The creation of patterned layers was not required for the suggested application but could, in principle, be

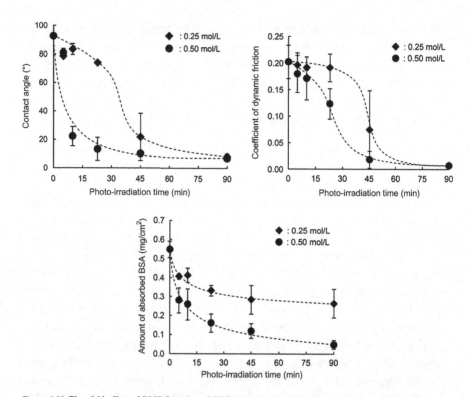

Figure 4.11 **Threefold effect of PMPC graft on PEEK: reduction in water contact angle, coefficient of dynamic friction, and BSA adsorption.** *Diamonds: monomer concentration 0.25 mol/L; circles: monomer concentration 0.5 mol/L. Source:* Adapted from Kyomoto *et al.* [22], with permission from Elsevier Ltd.

implemented by masking part of the substrate during the UV-induced polymerization step.

The findings of the study by Kyomoto *et al.* [22] have been extended to include carbon-fiber-reinforced PEEK as a substrate material [23]. Moreover, electron spin resonance was used to characterize radical formation upon irradiation with UV light, indicating that a steady state in radical creation was already reached after 15 min of illumination. Again, graft layers of PMPC were produced with the aim to improve the durability of an artificial hip. The gravimetric wear tested under realistic conditions was significantly decreased for PMPC-modified PEEK surfaces, whereas the bulk mechanical properties of the material were shown to be unaffected.

Polyurethanes (PUs), another technologically relevant class of polymers frequently used in blood- and other fluid-containing devices, have

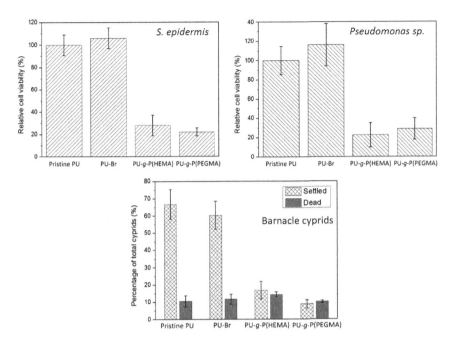

Figure 4.12 **Effect of grafting PHEMA and PPEGMA on PU surfaces.** (Top) *Reduced microfouling—that is, bacterial cell viability on surface grafted with PHEMA and PPEGMA compared to pristine PU and PU with covalently bound macroinitiator (PU-Br).* (Bottom) *Reduced macrofouling—that is, reduction in the amount of settled, live barnacle cyprids. Source*: Adapted from Pranantyo *et al.* [24], with permission from RSC Publishing.

been modified to reduce fouling [24]. A bifunctional macroinitiator containing an azide moiety and an ATRP initiator was synthesized in a multistep procedure. The azide was used to attach the macroinitiator to the PU substrate via nitrene insertion reaction, whereas the ATRP initiator was used in the subsequent controlled polymerization of 2-hydroxyethylmethacrylate (HEMA) or PEGMA. Micropatterns of these brushes were prepared by irradiation through a 200-mesh TEM grid during the macroinitiator immobilization step, resulting in feature sizes of approximately 80 μm. The polymer brushes were shown to significantly reduce microfouling—that is, adhesion of *Staphylococcus epidermis* and *Pseudomonas sp.*, as well as adhesion of barnacle cyprid, which is one of the most prominent macrofoulers in the marine environment (Figure 4.12).

Our group has studied different strategies for creating biofunctional brushes on ETFE. Applicable structuring methods include EUV-interference lithography for nanopatterns or masking (e.g., with TEM grids) during plasma activation for micropatterning. As illustrated in

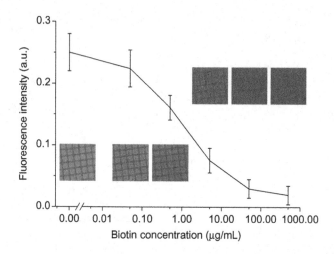

Figure 4.13 Competitive fluorescence assay performed on arrays of 40-μm-sized squares of PGMA structures grafted from ETFE. The samples were post-modified with biotin and incubated with fluorescently labeled streptavidin and biotin of increasing concentrations (0, 0.005, 0.05, 0.5, 5, and 50 μg/mL). Relative fluorescence intensities were determined from fluorescence images shown as insets.

Figure 4.1, poly(glycidyl methacrylate) serves as a starting point for different post-modification strategies. Microstructured biotinylated brushes were prepared and challenged with solutions containing the fluorescently labeled conjugate streptavidin and increasing concentrations of biotin (Figure 4.13) [4].

So far, we have summarized strategies to exploit the chemical versatility of polymer brushes to either immobilize biomolecules by covalent attachment or for significantly decreasing protein adsorption. However, the "extended interface" created by the brush in a good solvent also provides a swellable, soft layer that can promote the nonspecific immobilization of enzymes and provide an environment that supports their activity. We have tested the functionality of enzymes physisorbed from solution [11]. Because this type of binding is weak, the conformation and activity of the proteins is expected to remain largely intact. To assess the influence of polymer brush chemistry, wettability, and swellability on the physisorption of proteins, model enzymes were chosen. Alkaline phosphatase (ALP) and horseradish peroxidase (HRP) were selected because they both catalyze the transformation of a colorless substrate to a colored product, and the enzymatic activity can therefore be easily monitored with colorimetry. The substrate of choice for ALP is *para*-nitrophenyl phosphate (pNPP), which is hydrolyzed to yield yellow *para*-nitrophenol (pNP) (Figure 4.14).

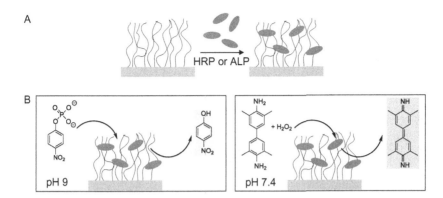

*Figure 4.14 **Polymer brushes with enzymatic activity.** (A) Enzymes are physisorbed in polymer brushes with different character (e.g., neutral or charged). (B) The activity of the enzymes such as alkaline phosphatase (left) and horseradish peroxidase (right) is assayed with colorimetry.*

HRP oxidizes 3,3′,5,5′-tetramethylbenzidine (TMB) in the presence of hydrogen peroxide to give a product with deep blue color. The enzymes differ significantly in their size and isoelectric point (IEP), with HRP being smaller (44 kDa) and having a higher IEP (7.2) compared to ALP (57 kDa, IEP 4.5).

Quantitative comparison of enzymatic activity is possible if the polymer brush layers are of the same thickness and have, in principle, the same capacity for protein physisorption. We therefore compared neutral poly(dimethylaminoethyl methacrylate) (PDMAEMA) and poly(4-vinylpyridine) (P4VP) brushes with their quaternized strong polyelectrolyte counterparts. Strikingly, enzyme activities in quaternized brushes were very high, whereas the neutral brushes exhibited no or only very low enzymatic activity (Figure 4.15). Although electrostatic attraction between the brushes and the enzymes might contribute to the physisorption of ALP and HRP, we also attribute the described behavior to the excellent water solubility of brushes with quaternary amines. These brushes were observed to swell drastically in aqueous solutions, suggesting that the capacity for physisorption of enzymes is very large.

Poly(acrylic acid) (PAA) brushes are also strongly swellable in water, particularly at neutral to basic pH, at which the carboxylic acid moieties are dissociated. Micro- and nanostructured brushes of PAA were prepared on ETFE foils using EUV interference lithography for activation. Microperoxidase 11 (MP11) is a heme-containing undecapeptide produced

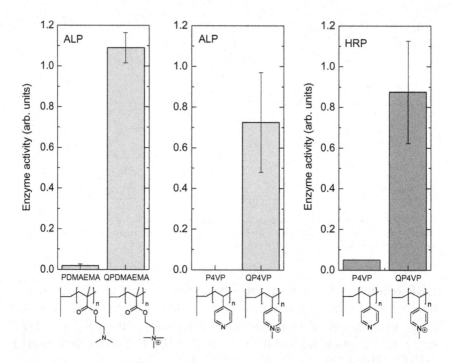

Figure 4.15 *Enzymatic activity of ALP* (left) *and HRP* (right) *immobilized in polymer brushes as assessed with colorimetry.* Very high enzyme activity was observed in the positively charged strong polyelectrolyte brushes, whereas enzyme activity was low in neutral brushes.

by enzymatic digestion of cytochrome *c* that exhibits peroxidase activity. MP11 was physisorbed in structured PAA brushes, and its activity was tested similar to the strategy discussed previously for HRP. The enzymatic reaction resulted in a blue self-staining of the brush. The intensity of the staining depended on the brush thickness, clearly indicating that the enzyme was physisorbed *into* and not only onto the brush (Figure 4.16).

These examples demonstrate how enzymatic activity can be locally introduced on surfaces using polymer brush structures and how their activity is influenced by both chemical functionality and thickness of the brushes.

4.4 CONCLUSIONS

Grafted polymer layers offer enormous possibilities to adapt and tune surface properties of polymers and to create stimuli-responsive or bio-functional surfaces. By appropriately choosing the grafting strategy

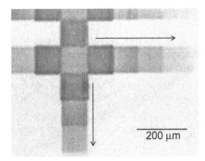

*Figure 4.16 **Structured PAA brush after physisorption of MP11 and enzymatic reaction.** The edge length of the individual squares is 100 μm. The brush is stained blue by the reaction products. The staining is clearly more intense in areas of thicker brushes. Arrows indicate decreasing brush thickness.*

(direct grafting vs. post-functionalization), a very broad range of chemical functionalities is accessible. Patterned functional surfaces enable local changes in surface properties or site-selective immobilization of biomolecules. The main challenge for the future is to transfer these functional surfaces from the lab bench to real-world applications.

REFERENCES

[1] Neuhaus S, Padeste C, Solak HH, Spencer ND. Functionalization of fluoropolymer surfaces with nanopatterned polyelectrolyte brushes. Polymer 2010;51(18):4037−43.

[2] Neuhaus S, Padeste C, Spencer ND. Functionalization of fluropolymers and polyolefins via grafting of polyelectrolyte brushes from atmospheric-pressure plasma activated surfaces. Plasma Process Polym 2011;8(6):512−22.

[3] Shkolnik S, Behar D. Radiation-induced grafting of sulfonates on polyethylene. J Appl Polym Sci 1982;27(6):2189−96.

[4] Padeste C, Farquet P, Potzner C, Solak HH. Nanostructured bio-functional polymer brushes. J Biomater Sci Polym Ed 2006;17(11):1285−300.

[5] Biesalski M, Johannsmann D, Ruhe J. Synthesis and swelling behavior of a weak polyacid brush. J Chem Phys 2002;117(10):4988−94.

[6] Zhou F, Huck WTS. Three-stage switching of surface wetting using phosphate-bearing polymer brushes. Chem Commun 2005;48:5999−6001.

[7] Heinz P, Bretagnol F, Mannelli I, Sirghi L, Valsesia A, Ceccone G, et al. Poly(N-isopropylacrylamide) grafted on plasma-activated poly(ethylene oxide): thermal response and interaction with proteins. Langmuir 2008;24(12):6166−75.

[8] Farquet P, Padeste C, Solak HH, Gursel SA, Scherer GG, Wokaun A. EUV lithographic radiation grafting of thermo-responsive hydrogel nanostructures. Nucl Instrum Methods Phys Res B 2007;265(1):187−92.

[9] Eaton P, West P. Atomic Force Microscopy. Oxford: Oxford University Press; 2010.

[10] Choi WS, Koo HY, Kim JY, Huck WTS. Collective behavior of magnetic nanoparticles in polyelectrolyte brushes. Adv Mater 2008;20(23):4504−8.

[11] Neuhaus S. Functionalization of polymer surfaces with polyelectrolyte brushes [dissertation]. Zurich: ETH Zurich; 2011.

[12] Breulmann M, Colfen H, Hentze HP, Antonietti M, Walsh D, Mann S. Elastic magnets: template-controlled mineralization of iron oxide colloids in a sponge-like gel matrix. Adv Mater 1998;10(3):237−41.

[13] Dante S, Hou ZZ, Risbud S, Stroeve P. Nucleation of iron oxy-hydroxide nanoparticles by layer-by-layer polyionic assemblies. Langmuir 1999;15(6):2176−82.

[14] Moya S, Azzaroni O, Farhan T, Osborne VL, Huck WTS. Locking and unlocking of polyelectrolyte brushes: toward the fabrication of chemically controlled nanoactuators. Angew Chem Int Ed 2005;44(29):4578−81.

[15] Azzaroni O, Brown AA, Huck WTS. Tunable wettability by clicking into polyelectrolyte brushes. Adv Mater 2007;19(1):151−4.

[16] Neuhaus S, Padeste C, Spencer ND. Versatile wettability gradients prepared by chemical modification of polymer brushes on polymer foils. Langmuir 2011;27(11):6855−61.

[17] Joseph G, Pichardo J, Chen G. Reversible photo-/thermoresponsive structured polymer surfaces modified with a spirobenzopyran-containing copolymer for tunable wettability. Analyst 2010;135(9):2303−8.

[18] Duebner M, Spencer ND, Padeste C. Light-responsive polymer surfaces via post-polymerization modification of grafted polymer-brush structures. Langmuir 2014;30(49): 14971−81.

[19] Yun JM, Jung CH, Kim DK, Hwang IT, Choi JH, Ganesan R, et al. Photosensitive polymer brushes grafted onto PTFE film surface for micropatterning of proteins. J Mater Chem 2010;20(10):2007−12.

[20] Fristrup CJ, Jankova K, Eskimergen R, Bukrinsky JT, Hvilsted S. Protein repellent hydrophilic grafts prepared by surface-initiated atom transfer radical polymerization from polypropylene. Polym Chem 2012;3(1):198−203.

[21] Coad BR, Lu Y, Meagher L. A substrate-independent method for surface grafting polymer layers by atom transfer radical polymerization: reduction of protein adsorption. Acta Biomater 2012;8(2):608−18.

[22] Kyomoto M, Moro T, Takatori Y, Kawaguchi H, Nakamura K, Ishihara K. Self-initiated surface grafting with poly(2-methacryloyloxyethyl phosphorylcholine) on poly(ether-ether-ketone). Biomaterials 2010;31(6):1017−24.

[23] Kyomoto M, Moro T, Yamane S, Hashimoto M, Takatori Y, Ishihara K. Poly(ether-ether-ketone) orthopedic bearing surface modified by self-initiated surface grafting of poly(2-methacryloyloxyethyl phosphorylcholine). Biomaterials 2013;34(32):7829−39.

[24] Pranantyo D, Xu LQ, Neoh KG, Kang ET, Yang W, Teo SLM. Photoinduced anchoring and micropatterning of macroinitiators on polyurethane surfaces for graft polymerization of antifouling brush coatings. J Mater Chem B 2014;2(4):398−408.

CHAPTER 5

Characterization Challenges of Micro- and Nanografted Polymer Systems

This chapter discusses challenges with the application of analytical tools for the characterization of polymer structures on grafted polymeric substrates. The chapter presents strategies to overcome these challenges, discusses the advantages and disadvantages of different analytical methods, and provides selected experimental results and illustrative case studies from the authors' experience and from the literature. It identifies a number of complementary characterization methods that are well suited for the investigation of polymer-on-polymer systems from the authors' point of view; however, it is not the aim of this chapter to provide a comprehensive overview of all available analytical methods or a thorough review of literature results. For details on the theoretical background and on the working principles of the methods, refer to the plethora of specialized books and articles available.

5.1 INTRODUCTION

Challenges in the characterization of micro- and nanopatterned polymer grafts on polymer supports originate from the following factors:

- *Grafted structures are small*: With increasing resolution, acquiring spatially resolved information on topography and chemical composition is increasingly demanding.
- *Surface grafting leads to thin films*: The graft layers obtained by micro- and nanografting usually have a thickness in the range of tens of nanometers up to several micrometers. These layers are extremely thin compared to the bulk polymer on which they are grafted. Obtaining quantitative information on these thin layers is very difficult because established polymer characterization methods probe the bulk of the polymer rather than its surface.
- *Polymers are grafted on polymers*: Basically all polymer-on-polymer systems are insulators. The use of state-of-the-art surface analysis

Polymer Micro- and Nanografting. DOI: http://dx.doi.org/10.1016/B978-0-323-35322-9.00005-X

tools (e.g., atomic force microscopy (AFM), scanning electron microscopy (SEM), and x-ray photoelectron spectroscopy (XPS)) is rendered more complicated by undesired charging, and in some cases, the achievable precision and resolution is reduced. In addition, base polymers and grafted polymers often have very similar chemical compositions, leading to low contrast between the two materials in many methods.

- *Experimental conditions*: Often, the environment (liquid/dry, ionic strength, pH, temperature, etc.) strongly influences the properties of the sample (see Chapter 4). However, many characterization techniques are limited to vacuum or air environment or difficult to apply in liquid.

Due to these difficulties, the combination of different methods is mostly needed to collect the desired information on the sample. The methods discussed here are divided into the subgroups of spectroscopic methods, surface analysis techniques, and microscopic techniques.

5.2 SPECTROSCOPIC METHODS

5.2.1 Infrared Spectroscopy

Infrared (IR) spectroscopy is a very attractive method for identifying chemical functionalities. Individual bonds between atoms (e.g., the double bond in $C = O$) or whole groups of atoms (e.g., in a CH_3-endgroup) are excited through absorption of IR radiation of a given wavelength, leading to vibration of different modes. As a consequence, the presence of functional groups such as double bonds, carboxylic acids, amines, amides, or hydroxyls results in distinct IR absorption bands.

Thus, the success of a grafting reaction can be qualitatively or even quantitatively assessed if the graft polymer exhibits characteristic IR bands that are different from those of the base polymer. However, only a thin layer of graft polymer is available for acquiring information, and the relevant signals can easily be obscured due to the dominance of the substrate material. This is illustrated in the top panel of Figure 5.1, in which a transmission spectrum of a 100-µm-thick poly(ethylene-*alt*-tetrafluoroethylene) (ETFE) foil with a poly(4-vinylpyridine) (P4VP) brush is shown. Clearly, the substrate bands are strong enough to completely block light transmission in a wide wavenumber range—that is, from 1350 to 950 cm^{-1}. This problem can be

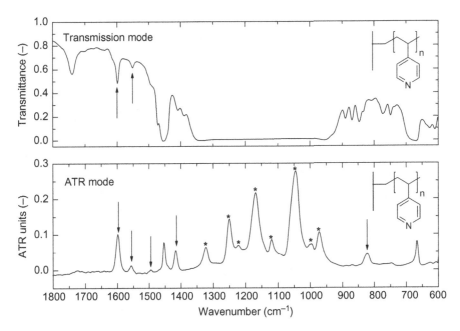

Figure 5.1 Comparison of IR spectra of a P4VP brush on ETFE recorded with an IR microscope in transmission (top) *and ATR mode* (bottom). *The contribution of the ETFE substrate was greatly reduced in ATR mode* (*ETFE bands marked with* asterisks*), and the bands corresponding to pyridine ring skeleton vibrations (marked with* arrows) *could be clearly identified.*

remedied by using attenuated total reflectance (ATR)-IR spectroscopy, in which a single crystal of high refractive index is pressed against the sample surface. An evanescent field of IR light is created that penetrates only the sample's surface. The penetration depth is dependent on the refractive indices of the crystal (e.g., germanium) and the sample, the wavelength, and the angle of incidence. For IR radiation, it typically lies in the range of approximately 0.5 μm. A spectrum recorded in ATR-IR mode is shown in the bottom panel of Figure 5.1. Again, an ETFE substrate with a P4VP brush was characterized, but in this case, the substrate bands are clearly resolved and additional bands from the graft polymer are identifiable.

If an *ATR-IR microscope* is used, the surface sensitivity of ATR-IR can be complemented with spatially resolved information. It allows collecting information in defined regions, and some instruments offer options for mapping the surface composition. The lateral resolution is limited by the crystal size and in the range of 100 μm. If no crystal is required and *IR microscopy* is sufficient for identification of relevant bands, the lateral resolution can be much lower. In this case, it is only determined by the wavelength of the radiation and the numerical aperture (NA) of the objective and is therefore in the range of $3-10\ \mu m$ for a good objective (NA ~ 0.6).

IR spectroscopy is well suited to monitor the progress of post-functionalization reactions—for example, for the quaternization of P4VP (see Figure 4.1B) or the hydrolysis or reduction of poly(vinylformamide) to yield primary amines (see Figure 4.1C) [1,2]. Moreover, the presence of undesired by-products such as boric acid after using borane/THF adducts for the aforementioned reduction reaction can be detected.

In another example, the ability of certain brushes to complex metal cations was investigated using ATR-IR spectroscopy. The potential of polymer brushes to act as polyligands for metal cations such as Fe^{2+} and Ni^{2+} is exploited, for instance, for the preparation of iron oxide-loaded brushes with magnetic properties (see Chapter 4) or for the creation of surfaces with a strong affinity to histidine-tagged proteins (discussed later). The successful loading of P4VP brushes with nickel cations was confirmed with ATR-IR spectroscopy, as illustrated in Figure 5.2. After complexation of Ni^{2+}, the intensity of the pyridine ring vibration band at $1598\ cm^{-1}$ was significantly reduced, and a shoulder appeared at $1615\ cm^{-1}$.

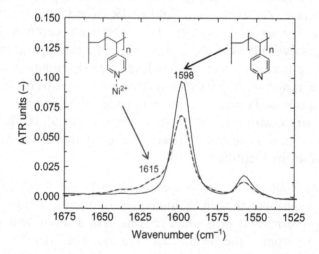

Figure 5.2 ATR-IR spectra of P4VP brushes in the as-grafted state (**solid line**) *and after complexation of Ni^{2+}* (**dashed line**). *A shoulder on the pyridine ring skeleton vibration at $1615\ cm^{-1}$ appears after Ni^{2+} complexation [3].*

5.2.2 Ultraviolet/Visual Spectroscopy

With ultraviolet/visible (UV/Vis) spectroscopy, a fast comparison of polymer foils in the pristine and modified state is possible. The samples must have a relatively high transparency for the light used for the analysis to be successful, which limits the number of applicable substrates and the maximum sample thickness.

Monitoring of graft modifications is possible if the graft polymer has a distinct absorption in the UV range. In Figure 5.3, UV spectra for a number of graft polymers on ETFE are shown [2]. Although the spectrum of poly(glycidylmethacrylate) (PGMA) brush-modified ETFE is very similar to the spectrum of the substrate, sulfonation of PGMA results in a clearly reduced transmittance up to 350 nm (PGMAS). For P4VP-modified ETFE, a strong decrease in transmittance below 250 nm was observed. This coincides with the absorption maximum of P4VP in solution caused by the heteroatomic ring. Quaternization of P4VP to yield QP4VP results in yet another significant change in the spectrum. In the case of poly(vinylformamide) (PVF) grafted from ETFE, the opacity below 250 nm can be attributed to the absorption by amides, which is typically strongest around 220 nm [4]. The presence of a very large number of identical functional groups within the thin brush layers accounts for the observed significant changes in the UV/Vis spectra.

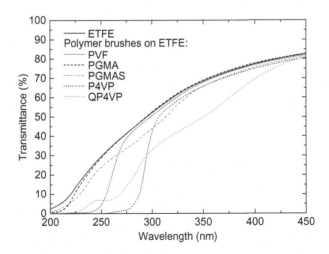

Figure 5.3 **UV spectra of different graft polymers on 100-μm-thick ETFE foils.** *See the text for details. Source*: Reproduced from Neuhaus *et al.* [2], with permission from John Wiley & Sons Inc.

As discussed in Chapter 4, UV/Vis spectroscopy can also be used to monitor the loading of polymer brushes with nanoparticles that absorb radiation in this wavelength range (see Figure 4.8).

Graft polymers loaded with dyes having strong absorption bands at a clearly defined wavelength are suitable for characterization with UV/ vis spectroscopy. Because extinction coefficients of dyes are very high, selective staining can aid in identifying and even quantifying functionalities in thin graft layers. We used an anionic dye, Coomassie Brilliant Blue (CBB), to monitor the time-dependent reaction of QP4VP with sodium hydroxide solution to yield a neutral poly[vinyl(N-methyl-2-pyridone)] (PVMP) brush (Figure 5.4) [5]. With increasing treatment time, the amount of positively charged N-methylvinylpyridinium groups decreased. As a consequence, the amount of CBB that could be loaded into the brush was also reduced.

The data presented here were acquired using nonstructured polymer films with large-area grafted polymer brush layers. Note that light can quite easily be focused to a spot size on the order of micrometers or be directed to the sample using fiber optics, allowing for the acquisition of data from areas with micrometer dimensions and for characterization of structured samples with micrometer resolution. Furthermore,

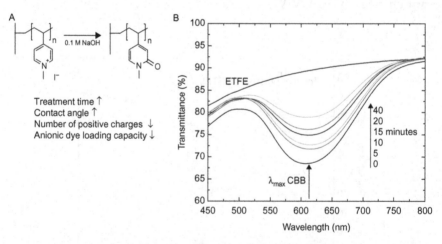

Figure 5.4 Reaction kinetics of a polymer brush modification traced with UV/vis spectroscopy. (A) Reaction of QP4VP to PVMP in sodium hydroxide. With increasing treatment time, the concentration of positive charges inside the brush decreases; thus, the loading capacity for Coomassie Brilliant Blue (CBB), an anionic dye, decreases as well. (B) Transmission spectrum of an ETFE foil grafted with Q4VP brushes and stained with CBB. The amount of CBB on the surface decreases with increasing treatment time. This allowed monitoring the conversion of QP4VP to PVMP [2]. A reference spectrum of a bare ETFE foil is shown for comparison.

selective staining enables easier identification of grafted structures using light microscopy.

5.2.3 Fluorescence Spectroscopy

In the previous example, CBB was loaded into a polymer brush to enable UV/vis analysis. Similarly, fluorescent molecules can be loaded or specifically attached to graft polymer-modified surfaces. Fluorescence signals are usually strong and specific and thus allow detecting small quantities of material. However, many polymers that could be used as base materials for grafting show autofluorescence or emit fluorescence due to the presence of additives. These signals may interfere with the specific signal of bound fluorescent moieties. The observation of the surfaces with fluorescence microscopy allows direct monitoring of grafted regions and at least qualitative comparison between fluorescence intensities in different regions of the sample or in graft layers of different thicknesses.

Fluorescence microscopy was used, for instance, to gain spatially resolved information on the modification of ETFE with a sulfonated brush. The brush was prepared by free radical polymerization following the irradiation of the substrate with EUV radiation in an interference setup. Fluorescently labeled poly(allylamine) was deposited on the brush-modified surface and specifically adsorbed on sulfonated regions by electrostatic interactions between sulfonate and amino groups [6]. Observation of the surface clearly showed increased fluorescence intensity in brush-modified regions (Figure 5.5).

Figure 5.5 **Fluorescent micrograph of a sulfonated polymer brush microstructure after adsorption of fluorescently labeled poly(allylamine).** *Source*: Reproduced from Padeste *et al.* [6], with permission from Taylor and Francis.

In another example, P4VP (see also Figure 5.2) or poly(vinylamine) brushes were complexed with Ni^{2+} cations and thereby prepared for specific immobilization of histidine-tagged proteins (Figure 5.6A). For proof of concept, histidine-tagged green fluorescent protein (GFP) was deposited on brushes loaded with Ni^{2+}. After incubation and extensive washing, the surface was analyzed using fluorescence microscopy [3]. For comparison, GFP was deposited on brushes without nickel (Figure 5.6B). The fluorescence intensity was clearly highest in regions modified with nickel-loaded brushes as shown in Figure 5.6C, indicating that a selective binding of histidine-tagged GFP had indeed taken place on brushes complexed with Ni^{2+}.

In more quantitative approaches, relative fluorescence intensities derived from fluorescence images were used to perform a competitive

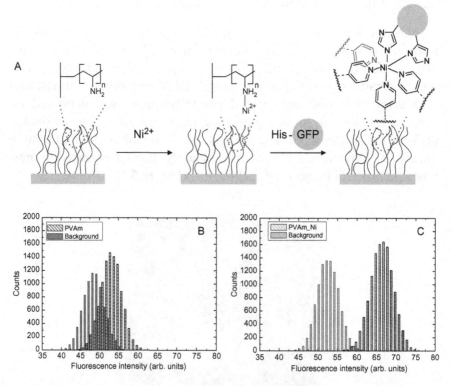

Figure 5.6 (A) A poly(vinylamine) brush is complexed with Ni^{2+}. Specific interactions between Ni^{2+} and histidine-tagged green fluorescent protein are then exploited for immobilization of the protein. (B) Fluorescence intensity of GFP on PVAm brush-modified regions (striped bars) and background intensity (solid bars) for comparison. (C) Fluorescence intensity of GFP on PVAm brushes loaded with Ni^{2+} (striped bars) and background intensity (solid bars) [3].

biotin/streptavidin assay (see Chapter 4, Section 4.3 and Figure 4.13) or to determine the switching kinetics of photoresponsive spiropyrans [7] on structured brushes grafted from ETFE surfaces.

5.3 SURFACE ANALYSIS TECHNIQUES

Surface analysis techniques specifically probe the surface of a material and are capable of detecting sub-monolayer surface coverage. However, charging in polymeric systems and the chemical similarity between the graft polymer and the substrate complicate the analysis in entirely polymeric systems.

5.3.1 X-Ray Photoelectron Spectroscopy

In XPS, the sample surface is exposed to monochromatic x-rays, whereby electrons are ejected due to the photoelectric effect. Their binding energy is calculated from the measured kinetic energy and used for determination of the elements present and their binding states. The probed depth is only a few nanometers, enabling an extremely surface-sensitive analysis of the material with a low detection limit (0.01 ng/cm^2). The lateral resolution is between 2 and 15 μm [8].

Charge compensation is usually necessary when investigating polymers to avoid charging and concomitant peak shifts and broadening. Elaborate peak fitting procedures are required to subtract contributions from the base material. Moreover, uncertainty in the quantitative interpretation of XPS spectra is caused by the presence of carbon contaminations on the surface, which complicate the interpretation of the C1s peak.

Lisboa *et al.* extensively exploited the possibilities of XPS analysis to study different graft polymer layers prepared by UV-induced free radical polymerization on poly(pyrrole) (PPy) films [9]. The aim was to prepare carboxylic acid or amine functionalized PPy for biomolecule immobilization. The contribution of the PPy carbon atoms in the C1s peak disappeared after a grafting time of 1200 s for acrylic acid and almost completely disappeared after 1500 s for allylamine, indicating that the PPy surface was completely covered with the graft polymer. For quantification of the amount of carboxylic acid and amine functionalities, a derivatization of the functional groups with trifluoroethanol and 4-trifluoromethylbenzaldehyde, respectively, was necessary.

Using this strategy, confusion of different contributions of COOX or CN signals was avoided, and relative quantification of functional groups in the graft polymer was straightforward by measuring the CF_3 peak area. Based on the results, the authors concluded that grafting of acrylic acid was far more efficient for PPy modification than grafting of allylamine.

The previous example illustrates that quantification on entirely polymeric systems is difficult to achieve and often requires elaborate post-modification steps. XPS analysis, however, can provide straightforward information on the presence of "hetero-atoms" from the graft polymer (e.g., phosphorus, nitrogen, and sulfur) if the substrate polymer does not contain these elements. Moreover, complexation of metal cations such as Fe^{2+}, Ni^{2+}, and Cu^{2+} can also be investigated. XPS is therefore complementary to other analysis techniques, such as UV spectroscopy and ATR-IR spectroscopy.

5.3.2 Time-of-Flight Secondary Ion Mass Spectrometry

In secondary ion mass spectrometry (SIMS), the sample is bombarded with primary ions resulting in sputtering of the surface. The ejected secondary ions have characteristic ratios of mass to charge (m/z) through which they can be identified and assigned, for example, to characteristic fragments of the different polymers. The detector in time-of-flight (TOF)-SIMS is based on the extraction of the secondary ions, which are then accelerated over a defined distance. The mass is calculated by measuring the ion's time of flight. A depth of several nanometers is probed with an ultralow detection limit (picograms to femtograms), but the analysis is only semiquantitative. Due to this very high sensitivity, SIMS measurements can usually be run in the so-called static mode. In this mode, the amount of sputtered material is so low that the composition is considered not to be influenced by the measurement. Lateral resolution of TOF-SIMS is on the order of 80–200 nm, which allows high-resolution mapping [8]. Sputter profiles are often used to gain depth-resolved information on the composition of the sample. However, charging of the surface often complicates the analysis by reducing both mass and spatial resolution.

The grafting of poly(acrylic acid) and poly(allylamine) from the previous example was also conducted in the presence of a metallic mask

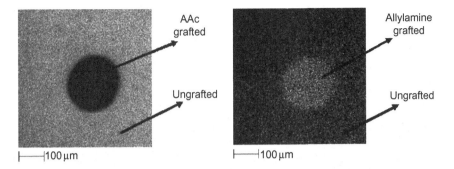

Figure 5.7 (Left) PPy with patterned poly(acrylic) acid graft. TOF-SIMS image of CN⁻ (m/z 26) ion distribution. CN⁻ ions originate from the substrate. (Right) PPy with patterned poly(allylamine) graft. TOF-SIMS image of C₃H₄NH₂⁺ (m/z 56) ion distribution. C₃H₄NH₂⁺ ions originate from the graft polymer. Source: Reproduced from Lisboa et al. [9], with permission from Elsevier Ltd.

with 150 µm circular holes. TOF-SIMS imaging was used to image the chemical contrast in grafted and ungrafted regions and therefore to demonstrate the precision and selectivity of the patterning process (Figure 5.7) [9]. In the case of poly(acrylic acid) graft polymers, CN⁻ ions originating from the PPy substrate were chosen for mapping, whereas ions ejected from the graft polymer were chosen in the case of poly(allylamine) modification. Both maps showed a distinct contrast between grafted and ungrafted regions and confirmed the circular pattern with 150 µm size.

5.4 MICROSCOPIC TECHNIQUES

5.4.1 Atomic Force Microscopy

AFM provides a wealth of possibilities for surface characterization. Inherent strengths of this method are its high resolution and the possibility to image in air or liquids without elaborate preparation steps. A particular challenge when dealing with grafted polymer substrates is the low sample conductivity, which leads to undesired charging effects. Moreover, the softness of the substrate renders stable imaging more difficult than on rigid substrates. Inherently, no two AFM tips will be the same; thus, variations in image quality or in quantitative analysis are unavoidable. For some applications, such as friction or adhesion measurements, it is advisable to use so-called colloidal probes, where a colloidal (silica-)sphere is attached to the end of the cantilever, which provides a much better defined contact zone [10,11]. Next, examples for some of the analysis modes of AFM are presented.

Topography Imaging in Non-Contact Mode

Soft structures such as polymer brushes are preferably probed in a non-contact AFM mode (e.g., tapping mode) to avoid ploughing through the material with the sharp AFM tip. An example of the topography of nanopatterned lines with 500-nm period is given in Figure 5.8. Using reversible addition fragmentation transfer polymerization on ETFE substrates activated with extreme ultraviolet (EUV) irradiation in an interference setup, poly(acrylic acid) (PAA) brushes and poly(acrylic acid-*block*-*N*-isopropylacrylamide) (PAA-*block*-PNIPAAm) brushes were prepared [12]. The images of the sample after the first grafting step demonstrate the high quality of nanostructures achievable using controlled polymerization schemes. In the second step, the structure height was clearly enhanced by the additional grafting of the PNIPAAm block.

Note that for *micropatterned* brushes, profilometry can deliver height profiles in a more straightforward and much less time-consuming manner. Because it is a contact method, lateral resolution is limited by the diameter of the profiler tips, which is typically in the range of 25 μm (smaller tips are available).

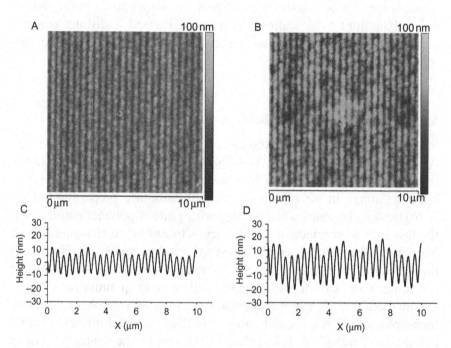

*Figure 5.8 **AFM images of 500 nm line patterns of polymer brushes on ETFE.** (A) Topography of PAA brushes and (B) topography of PAA-block-PNIPAAm brushes. (C) and (D) Average height profiles corresponding to (A) and (B), respectively. Source: Reproduced from Farquet et al. [12], with permission from ACS.*

Amplitude and Phase Shift Signals in Non-Contact Mode

In some cases, the amplitude or phase shift signals can provide additional information and/or better contrast than the topography mode. The shape of the structures may appear more distinct using *amplitude* images because this corresponds to an image of the *slope* of the features on the sample [13]. As discussed in Chapter 4 and shown in Figure 4.6, *phase* images can illustrate changes in viscoelastic surface properties—for example, changing softness when the sample is heated above the lower critical solution temperature (LCST). Generally, the phase signal is influenced by a number of factors, all of which contribute to the energy dissipation between the tip and the sample [13].

Adhesion Measurements

Adhesion between tip and sample can be probed by contacting the surface with the AFM tip and then measuring the force–distance curve upon retraction of the tip. Although this measurement mode is routinely available in AFMs, a number of potential pitfalls must be considered before evaluating the data. First, to avoid the influence of capillary forces between sample and tip, measurements in liquid are preferable. Second, tip geometry can vary substantially from tip to tip, leading to a large variation in contact area. Measurements can be compared only if the same tip was used throughout. To avoid this problem, a probe with defined spherical geometry (i.e., a colloidal probe) may be used. Third, the surface chemistry of the tip must also be considered. For example, the sign of the surface charge of the tip switches when changing pH across the isoelectric point of the material. Moreover, measurements at very basic pH (>9) are detrimental for SiO_2-covered tips because the material dissolves quickly under these conditions. Tips with defined surface functional groups were prepared, for instance, by depositing CH_3-, COOH-, and NH_2-terminated self-assembled monolayers on AFM tips [14,15]. A representative example for adhesion measurements is given in Figure 5.9. Below the LCST, the PNIPAAm brush showed no adhesion to the tip, whereas a measurable adhesion force was found above the LCST—that is, in the collapsed state of the brush [16].

Friction Measurements

Friction measurements are performed by measuring the twisting of the cantilever in AFM contact mode—that is, the lateral deflection. Stronger deflection is observed for higher friction between sample and

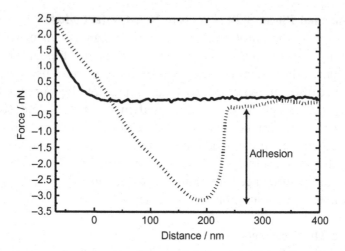

Figure 5.9 ***Adhesion between Si₃N₄ AFM tip and a PNIPAAm brush probed in water (18 MΩcm).*** *The curves illustrate the reversible temperature-dependent adhesion switching shown by PNIPAAm brushes. (*Solid line*) At 28° C (<LCST), there are no measurable adhesion forces. (*Dashed line*) At 40° C (>LCST), adhesion force is approximately 2.25 nN [16].* Source: Reproduced from Jones *et al.* [16], with permission from John Wiley & Sons Inc.

tip. However, one needs to take into account that topography can also contribute to the lateral deflection signal [13]. To achieve good reliability and reproducibility of friction measurements, the same considerations as for the adhesion measurements discussed previously apply (medium, tip chemistry, and geometry). We found that *qualitative* results can be quite easily interpreted. However, obtaining reliable *quantitative* data proved to be a major challenge, mostly because of the large variation in geometry and force constant from tip to tip but also because a large number of parameters influence the measurement. To illustrate a successful qualitative characterization of pH-dependent friction, height and friction images of poly(methacrylic acid) (PMAA) brush structures are shown in Figure 5.10 [3]. Images in Figure 5.10A were recorded in water at pH 3, at which the brush is in the neutral (i.e., protonated) state. In both height and friction images, the 283-nm dot pattern is clearly resolved, even though the quality of the images at higher normal force (24 nN) is better. In contrast, at pH 8 (Figure 5.10B), the brush is negatively charged. Because of the electrostatic repulsion between tip (SiO₂) and brush, friction values were very low. At low normal force, the tip could not be forced in close enough contact with the surface to allow for proper imaging. Higher normal forces were necessary to resolve the pattern.

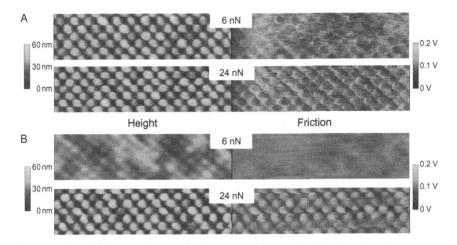

*Figure 5.10 **AFM images of a nanopatterned PMAA brush in height** (left) **and friction** (right) **mode.** Images were acquired in liquid with the normal force set to the indicated value. The scans are 5 × 1.25 μm. The solutions were at (A) pH 3 and (B) pH 8. A sharp silicon tip with a native SiO$_2$ layer was used [3].*

AFM Characterization of Soft Samples: Recent Trends and Developments

AFM technology is under intense development, particularly for the characterization of soft samples including biomaterials such as cells or proteins, as well as polymers and polymer composites. One major focus of these development efforts is the simultaneous quantitative mapping of topography and material properties such as elasticity, electrostatic and adhesive properties, and local energy dissipation. In addition to so-called quasi-static techniques, in which force-distance curves are acquired sequentially point per point on the whole surface, various approaches are followed to develop dynamic techniques. This development aims at the extraction of information similar to that of force-distance curves from the influence of the surface on the oscillation of a cantilever [17]. Operating principles involve, for instance, the simultaneous excitation and detection of two cantilever eigenmodes [18] or applying an oscillation with a much lower frequency than that of the cantilever. The amplitude of this low-frequency oscillation forces the tip to jump in and out of contact with the surface in every cycle [19]. Such advanced modes were successfully employed, for example, to characterize thin films of polymer blends [18], and they are of great interest for the characterization of polymer structures on polymer surfaces.

5.4.2 Scanning Electron Microscopy

In principle, high-resolution images elucidating the morphology of micro- and nanopatterned polymer grafts can be obtained with SEM. Depending on the detector used, information related to topography (secondary electrons) or composition (backscattered electrons) can be gathered. If energy-dispersive x-ray spectroscopy is available, the elemental composition can be determined in a given spot, line, or area.

When characterizing grafted polymer structures, the use of SEM is limited by the low conductivity of the polymer substrates. Coating of the samples with metals such as gold or chromium or with carbon is indispensable. However, the morphology of the coating can dominate the appearance of the surface, making the analysis prone to misinterpretations. A trade-off can be achieved by operating the SEM in "environmental" or "variable" pressure modes. Higher pressures maintained in the sample chamber of the instrument using differential pumping systems provide enough conductivity to equilibrate charging, eliminating the need to coat the substrates. However, resolution is not as good as in low-pressure modes. Further issues with SEM concern radiation damage and carbon deposition on the sample, which almost invariably occur after a given observation time. The severity of these effects depends strongly on beam intensity and electron acceleration voltage.

The difficulty of interpreting SEM images is illustrated in Figure 5.11. A P4VP brush structure was grafted from ETFE

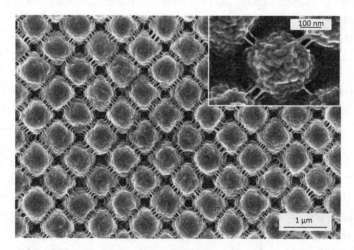

Figure 5.11 **SEM image of a dot pattern of P4VP brushes on ETFE (period 707 nm).** *The brush was loaded with iron cations and then aged. A chromium coating was applied to avoid charging.*

substrates activated with EUV light in an interference setup, resulting in a dot pattern with 707-nm period. The brush was loaded with an iron salt and then subjected to oxidation and aging. The aim of the SEM analysis was to determine whether iron oxide nanoparticles had been formed. To avoid charging, the sample was coated with chromium. The morphology of the surface is resolved down to the polymer "bundles" between individual dots. The cauliflower-like appearance of the dots is probably due to the collapse of the brush upon drying. However, it is not possible to judge whether the finest features within the dots have to be interpreted as chromium grains or as iron oxide nanoparticles.

5.4.3 Transmission Electron Microscopy

TEM provides the ultimate resolution necessary for characterizing features on the single-digit nanometer scale in polymer brushes. However, very elaborate sample preparation is required. Because samples are imaged in transmission, ultrathin cuts on the order of 100-nm thickness have to be prepared. For polymer samples, cuts can be prepared using a microtome or focused ion beam. By preparing transversal cuts, information on pattern fidelity and resolution is almost certainly lost. For a given sample thickness, contrast in TEM arises from mass contrast, as electrons interact more strongly with heavy than with light atoms. Consequently, contrast is very low in polymer samples with polymer grafts because the materials are composed of atoms of similar mass (mostly C, O, and N). This problem can be overcome by selective "staining" with metal cations. An example is given in Chapter 4 (see Figure 4.8), in which a P4VP brush was loaded with iron oxide nanoparticles resulting in a pronounced contrast in TEM images of cross sections of the sample.

5.4.4 Scanning Transmission X-Ray Microscopy

The scanning transmission x-ray microscopy (STXM) method requires a highly focused, extremely brilliant x-ray beam and is therefore confined to locations where synchrotron sources are available. It enables high-resolution imaging at different x-ray energies. The sample preparation is identical as for the TEM samples described previously. At the STXM of the PolLux beamline of the Swiss Light Source (SLS), the sample is rastered across the focus of the stationary x-ray beam by a movable stage [20,21]. The spatial resolution of the microscope is limited by the zone plate used for focusing and is currently in the range of 10–20 nm.

Element-specific contrast is obtained due to different absorption of the elements and can be enhanced by probing samples with photon energy slightly above the x-ray absorption edge of the element of interest. In the case of previously discussed P4VP brushes with Fe_xO_y nanoparticles, the sample strongly absorbs x-rays just above the iron edge (i.e., at 713.5 eV). In the regions where iron is present, the image appears dark, as shown in Figure 5.12. Below the iron edge (i.e., at 705 eV), the x-ray absorption of iron is much weaker. Therefore, regions with polymer brush/iron oxide composites can be clearly identified with high spatial resolution.

X-ray absorption spectra are obtained by scanning the same region multiple times with a succession of photon energies in the range of the absorption edge of a selected element. Evaluation of the absorption in the region of interest throughout the whole image stack yields absorption spectra. From the shape of the recorded spectra, further information such as the oxidation state [22,23] or ligand–metal interactions [24,25] can be derived. However, special care must be taken with polymeric samples because they are prone to beam damage during the long spectrum acquisition times (typically tens of minutes).

We obtained x-ray absorption spectra at the iron L_3 and L_2 edge from the P4VP brush/iron oxide composites discussed previously. In Figure 5.13, the spectrum of the P4VP brush/iron oxide composite of

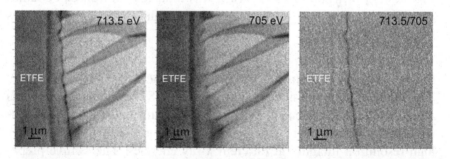

Figure 5.12 Scanning transmission x-ray micrographs obtained with a 15nm Fresnel zone plate on a microtome cut of an ETFE foil grafted with poly(4-vinylpyridine) (P4VP) brushes. The brushes were loaded three times with iron cations and then oxidized as illustrated in Figure 4.7. Images were recorded slightly above (713.5 eV; left) and below (705 eV; middle) the iron L_3 edge. To highlight the contrast caused by iron-containing particles, the two images were divided (right) [3].

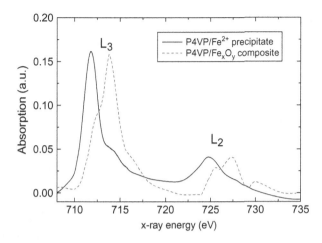

Figure 5.13 **X-ray absorption spectra at the iron L_3 and L_2 edges.** *(Solid line) P4VP/Fe^{2+} complex dropcast from solution. (Dashed line) Nanocomposite consisting of a P4VP brush and Fe$_x$O$_y$ [3].*

interest is shown as dashed line. For comparison, the spectrum of a model system consisting of a dropcast P4VP/Fe^{2+} precipitate was recorded (shown in black). At the L_3 edge, the two spectra differ significantly in the position of the absorption maximum and peak shape. At the L_2 edge, the peak of the composite is shifted to higher energy and split into two subpeaks. The strong differences in the spectra suggest that the transformation from a P4VP/Fe^{2+} complex to the P4VP/iron oxide nanoparticle composite was successful. Comparison of the P4VP/Fe$_x$O$_y$ spectrum with literature data [26] indicated the presence of both γ-Fe$_2$O$_3$ (maghemite) and Fe$_3$O$_4$ (magnetite).

5.4.5 X-Ray Microtomography

As discussed previously, elaborate and often tedious preparation steps are required for samples to be imaged with high-resolution transmission microscopy techniques such as TEM and STXM. Even if performed with great care, the preparation of ultrathin cross sections always bears the risk of influencing the sample's morphology by the cutting process. In particular for micrografted samples, synchrotron-based microtomography provides a valuable alternative because millimeter-sized objects can be probed in 3D with high spatial resolution reaching the sub-micrometer range [27].

In x-ray microtomography, the sample is mounted on a rotation stage and exposed to a synchrotron x-ray beam. Absorption images are then recorded using an x-ray pixel detector. From projection images acquired in all directions, a 3D map of the absorption in the sample is mathematically reconstructed. The data are usually represented as a series of 2D slices or as a 3D rendering of, for instance, the different phases of a composite material.

Similar to STXM, the selection of the wavelength of the x-ray light can be used to gain element-specific information or to optimize the contrast between different components in a multiphase system. In addition to absorption mapping, the acquisition of phase contrast images has been established, for example, based on grating interferometry [28].

Figure 5.14 illustrates the application of x-ray microtomography to an ETFE-based sample, which was exposed at a LIGA beamline, grafted with styrene, and then sulfonated (see Chapter 2, Section 2.4). Data were acquired at the TOMCAT beamline at the SLS, which has been established as a highly automated high-throughput facility [29,30]. To increase the absorption contrast, cesium chloride was diffused into the grafted regions of the sample. The vertical slice through the sample (Figure 5.14A) demonstrates that the reaction was

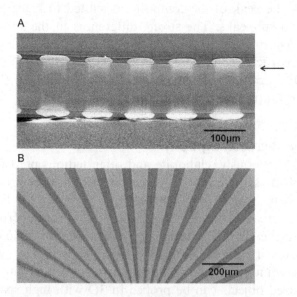

Figure 5.14 **X-ray absorption images of a bulk-grafted sample extracted from a micro-tomographic 3D data set.** *Styrene-grafted and sulfonated regions appear bright due to the increased absorption of x-rays by cesium ions diffused selectively into the structure. The two images represent two cross sections through the film: (A) perpendicular to the film and (B) parallel to the surface at the height indicated by the arrow in panel A. Without the need to cut the sample, fine features such as reaction fronts are clearly visible.*

terminated in this case before the grafting fronts from both sides of the film could meet in the center of the film. Grafting was found to proceed faster along the border between exposed and nonexposed areas, leading to a curved shape of the reaction fronts. Furthermore, the image indicates that under the chosen reaction conditions, grafting in the surface-near region of the polymer film was favored. The cross section parallel to the surface (Figure 5.14B) was extracted from the same data set and demonstrates the clear structure definition inside the film.

5.5 CONCLUSIONS

Advanced analytical methods enable an in-depth study of chemical and morphological characteristics of micro- and nanografted polymer materials. Often, the combination of several techniques is necessary to obtain a detailed picture. Investigations of simplified model systems such as nonstructured grafted surfaces, brushes produced on nonpolymeric substrates, or selectively derivatized or stained samples are needed for comparison or as standards. Furthermore, the adaptation of biochemical and chemical assay techniques is useful to complete the picture of the produced structures and their functionalities.

REFERENCES

[1] Neuhaus S, Padeste C, Solak HH, Spencer ND. Functionalization of fluoropolymer surfaces with nanopatterned polyelectrolyte brushes. Polymer 2010;51(18):4037−43.

[2] Neuhaus S, Padeste C, Spencer ND. Functionalization of fluropolymers and polyolefins via grafting of polyelectrolyte brushes from atmospheric-pressure plasma activated surfaces. Plasma Process Polym 2011;8(6):512−22.

[3] Neuhaus S. Functionalization of polymer surfaces with polyelectrolyte brushes [dissertation]. Zurich: ETH Zurich; 2011.

[4] Yamada B, Zetterlund PB. General chemistry of radical polymerizations. In: Matyjaszewski K, Davis TP, editors. Handbook of Radical Polymerization. Hoboken, NJ: Wiley; 2002.

[5] Neuhaus S, Padeste C, Spencer ND. Versatile wettability gradients prepared by chemical modification of polymer brushes on polymer foils. Langmuir 2011;27(11):6855−61.

[6] Padeste C, Farquet P, Potzner C, Solak HH. Nanostructured bio-functional polymer brushes. J Biomater Sci Polym Ed 2006;17(11):1285−300.

[7] Duebner M, Spencer ND, Padeste C. Light-responsive polymer surfaces via post-polymerization modification of grafted polymer-brush structures. Langmuir 2014;30 (49):14971−81.

[8] Mathieu HJ, Bergmann E, Gras R. Analyse et technologie des surfaces. Lausanne: Presses Polytechniques et Universitaires Romandes; 2003.

[9] Lisboa P, Gilliland D, Ceccone G, Valsesia A, Rossi F. Surface functionalisation of polypyrrole films using UV light induced radical activation. Appl Surf Sci 2006;252(13):4397−401.

[10] Ramakrishna SN, Nalam PC, Clasohm LY, Spencer ND. Study of adhesion and friction properties on a nanoparticle gradient surface: transition from JKR to DMT contact mechanics. Langmuir 2012;29(1):175–82.

[11] Nalam PC, Ramakrishna SN, Espinosa-Marzal RM, Spencer ND. Exploring lubrication regimes at the nanoscale: nanotribological characterization of silica and polymer brushes in viscous solvents. Langmuir 2013;29(32):10149–58.

[12] Farquet P, Padeste C, Solak HH, Gursel SA, Scherer GG, Wokaun A. Extreme UV radiation grafting of glycidyl methacrylate nanostructures onto fluoropolymer foils by RAFT-mediated polymerization. Macromolecules 2008;41(17):6309–16.

[13] Eaton P, West P. Atomic Force Microscopy. Oxford: Oxford University Press; 2010.

[14] Frisbie CD, Rozsnyai LF, Noy A, Wrighton MS, Lieber CM. Functional-group imaging by chemical force microscopy. Science 1994;265(5181):2071–4.

[15] Vezenov DV, Noy A, Rozsnyai LF, Lieber CM. Force titrations and ionization state sensitive imaging of functional groups in aqueous solutions by chemical force microscopy. J Am Chem Soc 1997;119(8):2006–15.

[16] Jones DM, Smith JR, Huck WTS, Alexander C. Variable adhesion of micropatterned thermoresponsive polymer brushes: AFM investigations of poly(N-isopropylacrylamide) brushes prepared by surface-initiated polymerizations. Adv Mater 2002;14(16):1130–4.

[17] Passeri D, Rossi M, Tamburri E, Terranova ML. Mechanical characterization of polymeric thin films by atomic force microscopy based techniques. Anal Bioanal Chem 2013;405 (5):1463–78.

[18] Garcia R, Proksch R. Nanomechanical mapping of soft matter by bimodal force microscopy. Eur Polym J 2013;49(8):1897–906.

[19] RosaZeiser A, Weilandt E, Hild S, Marti O. The simultaneous measurement of elastic, electrostatic and adhesive properties by scanning force microscopy: pulsed-force mode operation. Meas Sci Technol 1997;8(11):1333–8.

[20] Raabe J, Tzvetkov G, Flechsig U, Boege M, Jaggi A, Sarafimov B, et al. PolLux: a new facility for soft x-ray spectromicroscopy at the Swiss Light Source. Rev Sci Instrum 2008;79:11.

[21] Watts B, McNeill CR. Simultaneous surface and bulk imaging of polymer blends with x-ray spectromicroscopy. Macromol Rapid Commun 31(19):1706–12.

[22] Park J, Lee E, Hwang NM, Kang MS, Kim SC, Hwang Y, et al. One-nanometer-scale size-controlled synthesis of monodisperse magnetic iron oxide nanoparticles. Angew Chem Int Ed 2005;44(19):2872–7.

[23] Sun SH, Zeng H, Robinson DB, Raoux S, Rice PM, Wang SX, et al. Monodisperse MFe2O4 (M = Fe, Co, Mn) nanoparticles. J Am Chem Soc 2004;126(1):273–9.

[24] Hocking RK, Wasinger EC, de Groot FMF, Hodgson KO, Hedman B, Solomon EI, et al. Fe L-Edge XAS Studies of K4[Fe(CN)6] and K3[Fe(CN)6]: A Direct Probe of Back-Bonding. J Am Chem Soc 2006;128(32):10442–51.

[25] Hocking RK, George SD, Raymond KN, Hodgson KO, Hedman B, Solomon EI. Fe L-edge x-ray absorption spectroscopy determination of differential orbital covalency of siderophore model compounds: electronic structure contributions to high stability constants. J Am Chem Soc;132(11):4006–15.

[26] Park J, An KJ, Hwang YS, Park JG, Noh HJ, Kim JY, et al. Ultra-large-scale syntheses of monodisperse nanocrystals. Nat Mater 2004;3(12):891–5.

[27] Landis EN, Keane DT. X-ray microtomography. Mater Charact 2010;61(12):1305–16.

[28] McDonald SA, Marone F, Hintermuller C, Mikuljan G, David C, Pfeiffer F, et al. Advanced phase-contrast imaging using a grating interferometer. J Synchrotron Rad 2009;16:562−72.

[29] Mader K, Marone F, Hintermuller C, Mikuljan G, Isenegger A, Stampanoni M. High-throughput full-automatic synchrotron-based tomographic microscopy. J Synchrotron Rad 2011;18:117−24.

[30] Marone F. Hintermuller C. McDonald S. Abela R. Mikuljan G. Isenegger A. et al. X-ray tomographic microscopy at TOMCAT. Proc. SPIE 7078, Developments in X-Ray Tomography VI, 2008; 707822.

Printed in the United States
By Bookmasters